Transformative Participation for Socio-Ecological Sustainability

Around the CoOPLAGE pathways

Emeline Hassenforder and Nils Ferrand (eds)

Éditions Quæ

To cite this book:
Emeline Hassenforder, Nils Ferrand (eds), 2024. *Transformative Participation for Socio-Ecological Sustainability. Around the CoOPLAGE pathways*. Versailles, éditions Quæ, 270 p.

Translation and proofreading: Erika Ferrand Cooper
Cover page and graphical summary: Morgane Ganault

Chapters 2-6, 9-14, 16, 18, and inserts 1 and 3 of this book are translated from an issue of *Sciences Eaux & Territoires* (N°35, 2021) entitled in French:
Des démarches participatives pour penser ensemble la gestion de l'eau et des territoires.

Éditions Quæ
RD 10, 78026 Versailles Cedex
www.quae.com – www.quae-open.com

© Éditions Quæ, 2024

ISBN (PRINT): 978-2-7592-3919-1
ISBN (eBook): 978-2-7592-3921-4

ISBN (PDF): 978-2-7592-3920-7
ISSN : 1773-7923

Table of contents

SO AS YOU SEE ON SLIDE 123, THE SOLUTION IS OBVIOUS, THE TOPOGRAPHIC AND HYDRAULIC ANALYSIS REVEAL THAT...
AVAL RIVER PROJECT
Topographic analysis
HEY, STOP!
WE WOULD LIKE TO TRULY PARTICIPATE IN THIS PROJECT.
LOOK, THERE ARE WAYS!
« TRANSFORMATIVE PARTICIPATION FOR SOCIO-ECOLOGICAL SUSTAINABILITY »
IT'S A HANDBOOK...
LET'S SEE WHAT IT RECOMMENDS TO GET A REAL PARTICIPATORY DECISION IN OUR PROJECT...
PART. 1 : THE FOUNDATIONS
WE AGREE ON WHAT IT MEANS TO GET INVOLVED IN THE DECISION.
PART 2 : ENGINEERING AND EVALUATING
WHERE ARE WE HEADING ?
WE CHOOSE AND FOLLOW OUR PARTICIPATION FLOW!
PART 3 : PARTICIPATORY MODELLING
WE BUILD A MODEL OF OUR SITUATION AND USE IT TO EXPLORE ALTERNATIVE FUTURES.
PART 4 : PLANNING FOR TRANSFORMING
WE BUILD AND COMMIT TO A COMMON PLAN TO REACH OUR DESIRED FUTURE.
SOCIO-ECOLOGICAL SUSTAINABILITY
MO - ILLUSTRATIONS
LESBULLESDEMO@LILO.ORG

PART 2
ENGINEERING AND EVALUATING PARTICIPATORY DECISION PROCESSES

PART 3
PARTICIPATORY MODELLING TO SUPPORT DECISION AND CHANGE

PART 4
PLANNING FOR TRANSFORMING TOWARDS SOCIO-ECOLOGICAL SUSTAINABILITY

Introduction

Emeline Hassenforder and Nils Ferrand

▸▸ Underlying principles and posture of the book

Our world needs to adapt rapidly to the extreme conditions we have imposed on ourselves. Otherwise, the many prophecies of collapse might be fulfilled. The challenges of the Anthropocene drive us to reconsider and reengineer our ways of thinking, acting and living together and with our environment. However, most current trends are taking us in the wrong direction. This is particularly in terms of consumption and behavioural patterns, systems of financial control at the international level, extractive natural resources strategies, deepening inequalities, lack of effective democracies, the surge in conflicts and wars, distrust between many social actors, and much more. In such a dire situation, where should we—as humans, practitioners and scientists—focus our energy and agency for change?

– After the anthropocentric posture of the past decades, we need to reconsider the environment as a degraded common, and not as a permanent commodity.

– Individualism and competition promoted by liberalism should be replaced with solidarity and respect among humans, and with the other living species and entities.

– The diversity of human beings, specifically their perceptions and aspirations must be acknowledged by all as an asset for confronting the complexity of the situation, as well as a potential limitation requiring new cooperative practices.

– Top-down approaches to public decision-making where public policies are decided by leaders, driven by crowd and media prejudice, and accepted by the people, need to be transformed to revalue the contributions of all stakeholders, increasing the relevance of and commitment to public policies through co-construction with serious methods.

– Leaving a post-colonial North-South posture, we should foster South-North and South-South strategies.

– We should endorse gender-sensitive and indigenist visions of the situation and of the potential options for change instead of the dominant (masculinist) one.

Scientific research, through the Intergovernmental Panel on Climate Change (IPCC), the Intergovernmental Science-Policy Platform on Biodiversity and Ecosystem Services (IPBES) and all other targeted programs, seeks to impact policies, behaviours and socio-technical alternatives, but fails to significantly adjust the trajectories of socio-ecological systems. New scientific postures that seek *transformative actionable*

knowledge, built from and with deep interaction with stakeholders and communities can help to overcome these issues. Specifically:
– The detached forms of science that throw standalone academic insights and solutions from the lab, should be re-integrated conceptually and practically with the boundaries of its own data and models, uncertainty about impacts in implementation contexts, a responsive posture, and acceptance of controversies.
– Extractive forms of science that collect data from the field and from the people—as is the case in many citizen sciences projects—but are steered by scientists for the sake of discovery, might switch to an interactive and constructive approach, where questions, processes and their implementation are co-evolved with the concerned social groups in their environment.
– Greedy accumulation of data and knowledge should be questioned in regards to its actual contribution to societal change: use and impact in science, society and policy.
– From a disciplinary science, we should turn towards an undisciplined form of research, which is dynamically responsive to the greatest challenges we face.
– We should restore the central role of social sciences for its capacity to deal with the current failures of techno-solutionism, and cope with social change and governance.

This book, "Transformative participation for socio-ecological sustainability", does not hold the keys to revolutionary change. Based on 20 years of intervention research, coordinated and international, it instead aims at presenting experiences and approaches attempting to embody the above-mentioned principles, with their pros and cons. We hope it may help other researchers and practitioners in developing and implementing their own successful participatory pathways for the benefit of the socio-ecosystems, progressing a few steps forwards against the fate of collapse and towards a better world.

▸▸ Why this title?

Supporting people and societies in adapting to their most urgent socio-ecological challenges is the overall goal we endorse in this book. In this regard, socio-ecological sustainability is the overall objective that this book seeks to contribute to. Our assumption is that this objective cannot be achieved without the enhanced participation of all stakeholders (from citizens to policy-makers) in the decisions that affect our social-ecological systems. This means that the participation of the various stakeholders must climb Arnstein's (1969) ladder of participation, i.e. no longer simply informing participants, but building their capacities to decide, act and adapt autonomously (Castoriadis, 1975) towards the sustainability of our socio-ecological systems. Participants must thus acquire a threefold capacity to assess their own situation within a global system, to develop and integrate feasible action plans to tackle their problems, and to self-organise in order to engage and steer their own adaptation pathways.

In this sense, participation must be transformative. And we argue in this book that this transformation needs to be accompanied by approaches, methods and concrete feedbacks, insofar as the participatory processes involved comprise several decision and action steps, and address complex questions of socio-ecological sustainability.

Beyond citizen sciences, beyond top-down "acceptology", beyond non-engaging or manipulatory communication, such transformative research is a new frontier, as well as a candidate "must-do" in the social and political agenda. Within a wide international community of researchers aware of the urgency, and committed for "action on the ground" with and for the people, the "CoOPLAGE" group has designed, tested, gathered and coupled a specific suite of methods and tools, over 20 years in more than 30 countries. CoOPLAGE is the French acronym[1] for "Coupling Open and Participatory Tools to Let Actors Adapt for Environmental Management". The group's ambition is therefore to instrument this transformative participation in order to contribute to socio-ecological sustainability. The CoOPLAGE group is an interdisciplinary group of researchers and practitioners made up of researchers from the G-EAU joint research unit "Water Matters"[2] in Montpellier who have built the CoOPLAGE suite of tools over the years, and their field partners, whose decision-making needs have driven the construction of the CoOPLAGE tools.

Initially focused on water management, the experiences of the CoOPLAGE group have broadened to encompass issues around sustainable development, poverty, land use, governance and transition. They have been implemented with governments, environmental management institutions, non-governmental organisations, public agencies, citizens' groups, consulting firms and other researchers. The CoOPLAGE tools target various needs in participatory decision-making, including non-canonical ones like the co-engineering of the participation procedure and rules themselves, social justice principles, or self-designed protocols for social impact assessment. This wide set of experiences led to diverse "pathways", i.e. contextual adaptation, redesign, troubles and uncertainties, which helped improving CoOPLAGE and ultimately structured this book.

⇥ Inside or outside the book

In a nutshell, this book aims to give practitioners and researchers an overview of a coherent body of work and results on participatory decision processes aiming at socio-ecological sustainability, implemented in several countries. It covers topics ranging from co-design of participatory processes, to diagnostic, planning, and monitoring and evaluation of processes and impacts—with a common framework based on participatory modelling.
– This book deals with the participation of any stakeholder (citizens, representatives of associations, administrations, private companies, etc.)
– It addresses support to processes initiated in public policies. Emergent, bottom-up or protest participatory approaches (e.g. social movements, contested zones, etc.) are not addressed here. Most authors intervene under public commissioning. It may also support bottom-up dynamics but the initial trigger is often administrative.

---

1. CoOPLAGE: "Coupler des outils ouverts et participatifs pour laisser les acteurs s'adapter pour la gestion de l'environnement".
2. The G-EAU joint research unit "Water Matters" brings together researchers from a wide range of disciplines to work on a common research topic: water. We develop approaches and tools to understand and support sustainable water transformations. G-EAU is part of the ICIREWARD Unesco Centre for Water in Montpellier and engaged in the I-Site Excellence Program of the University of Montpellier. The academic and support staff of G-EAU involves the following institutions: the French National Research Institute for Agriculture, Food and Environment (INRAE), the French Agricultural Research Centre for International Development (Cirad), the French National Research Institute for Sustainable Development (IRD), AgroParisTech, Institut Agro Montpellier and the French Geological Survey (BRGM).

– It deals with participatory *processes*, i.e. including different steps and methods, multiple actors and issues, targeted at a specific change. It does not address *per se* group dynamic or facilitation techniques.

– It is focused on participatory decision-making and action support. It does not cover the lowest ladders of the Arnstein (1969) classification, like information, communication, consultation, and generally refutates "acceptology".

– It does not focus on science targeted participation or citizen sciences. Only one chapter addresses participatory observation (chapter 16).

– It mainly addresses physical or material-based processes, with in-person presence of participants. Only one chapter deals with digital participation (chapter 8) and mainly for its engineering and management. The book does not address electronic debate, pooling or online participatory budgeting.

▶▶ Key concepts and definitions

The definitions presented below are those used by the authors. Alternatives may exist in the literature.

– **Stakeholders**: all people or organisations affected by, or potentially affecting, the decision-making process (adapted from Glicken, 2000). e.g.: local authorities, non-governmental organisations (NGOs), companies, inhabitants, tourists, etc.

– **Citizens**: persons engaged in the "life in society", as a community of humans, and who holds some dedicated rights and duties. We restrict citizens to individuals and distinguish them from "representatives" of a civil group, company or any other organisation. e.g.: lay people, the locals, local population, the "general public", etc.

– **Participation**: involvement of stakeholders in decision-making or implementation processes from which they are usually absent, with various *intensity* from simple dialogue to co-management.

– **Participation engineering**: design and operational management of the participatory processes, by assessing context, needs, constraints, goals, and deciding participatory steps, participants, methods, regulation, and finally implementing it with adaptive steering.

– **Participatory**: variant of a given social or political process to its participatory form, with an inclusive approach for design and conduct. E.g. participatory modelling, participatory engineering, participatory observation, participatory monitoring...

– **Consultation**: The French word "concertation" can be roughly translated as consultation. "Concertation" in French is often used interchangeably with the word participation. We use the term consultation to designate participation including solely representatives of stakeholders (local authorities, associations, private companies) and not direct participation from citizens. An example of consultative body is local water committees.

– **Engagement**: action of becoming involved in or towards a participatory or decision-making process, with or for one or more other stakeholders. Engagement can be more or less deliberate (often referred to as involvement or commitment), or externally imposed (by a norm, contract, law, etc.). Disengagement, on the other hand, is the act of not getting involved (in a participatory or decision-making process in particular) and can be reflected in electoral abstention, a drop in associative participation or the weakening of trade union organisations (based on Luneau, 2013).

– **Socio-ecological sustainability**: for a coupled system where human communities interact with their surrounding ecological system (natural environment), the property of preserving the viability (existence and persistence of the state and functions) of the social and the ecological sub-systems, in short and longer term, under changing external constraints.

– **Governance**: effective decision-making processes in a given social system, combining formal rules and institutions, and informal but operative processes.

– **Environmental management**: tactical decisions and their implementation related to the preservation or restoration of the environment of a given social and economic system. May include public and private management, as well as individual behaviors as components of the effective management. Different from Governance which sets the strategic decision and the overall conditions of the management.

– **Decision process/decision cycle**: sequence of social interactions, sometimes structured by external interventions or methods, and leading to some actuated decisions, by persons and groups. A substrate of governance and management.

– **Autonomy**: conditions of a social group to self-decide its own goals and rules, and be able to follow them, without external interference or influence on alternatives, choice or implementation (based on Castoriadis, 1975). In a contemporary and materialistic form, property of a social group to be able to live without influence or dependencies from others, for instance by controlling its own metabolism for basic needs.

– **Modelling**: social process producing a model of a system (an intermediary or boundary object), i.e. a representation under some formalism (descriptive and explanatory language) which can help analysing and managing the same system (based on Minsky, 1965). Often restricted to specialists ("modelers"), it can be extended to participatory modelling where any stakeholder can take part, share her vision and "adopt" the resulting model. Such process is potentially transformative through the induced social learning.

– **Simulation**: activation of a model to assess (with or without a computer) some dynamics in response to initial situations, scenarios, inputs or triggers. Often used for testing management options. *Participatory simulation* (games and role-playing games) are specific types of simulations where some stakeholders "stay in the loop" of simulation, by observing and reacting dynamically to the evolution, around the table or through computers, to exhibit realistic decisions and behaviors. *Social simulations* are representing humans and dynamics of social groups, under various assumptions inspired from social sciences, in interaction with others and the environment.

– **Citizen sciences/participatory sciences**: engagement of citizens in the production of scientific knowledge, by asking them to observe and collect data (e.g. plants, animals in their environment), sometimes formulating analysis or questions.

– **Acceptology**: approach of governance and management where some decisions are pre-structured or pre-made by a group of policy makers and usually experts, who in a second step organise a limited participatory process aimed at getting these decisions to be accepted by other stakeholders, mainly citizens, expecting minimal contest and change of the pre-decision.

– **Participatory planning**: a decision process aimed at getting the participants to co-construct, adapt and adopt an action plan, i.e. a set of different tentative actions, organised in space and time, for one or many sectors or issues, with the constraint of ensuring its feasibility, efficiency and robustness in front of various scenarios.

– **Monitoring and evaluation (of participatory processes)**: a way to collect and provide useful data at the right time and in the right format to the actors who need it to make decisions towards socio-ecological sustainability. Participatory observatories can thus be seen as a perennial form of monitoring and evaluation of participatory approaches or of the socio-ecological systems in which they are rooted, aiming at providing reliable information to renew knowledge and support policy-making. *Monitoring* can be distinguished from evaluation in that it is a way of collecting and providing data throughout the process with the aim of improving and adapting it when necessary. *Evaluation* is more punctual (ex-ante, in-itinere, ex-post) and aims at assessing the value of the process (efficiency, impact, relevance, sustainability...) in order to provide relevant lessons for the upcoming or future processes. Monitoring is often done by people involved in the process while evaluation is often done by external people.

▸▸ Content of the book

This book includes an introduction composed of several chapters, and four parts. The current section lays the foundation for the book and draws the link among the various chapters. The introductory chapter 1 puts the content of the book into perspective in relation to what is being done elsewhere on the same subject. It specifies the values and postures underlying the approaches presented in the book, what these approaches are inspired by, and on the contrary, what they do not address. This perspective is at once historical, geographical, prospective and thematic. Chapter 2 presents the CoOPLAGE approach, its historical background and a set of complementary tools designed to meet the needs of stakeholders in supporting socio-environmental transition. The CoOPLAGE approach is in some ways the umbrella that embraces most of the chapters of this book: only six of the 19 chapters do not refer to the CoOPLAGE approach (the 3rd introductory chapter, as well as chapters 4, 6, 11, 15, 16). The introductory chapter 3 is a cross-talk between three people evoking the context of citizen participation in water management in France: a facilitator working for a non-governmental organisation who has been accompanying and facilitating local participatory processes for eight years, the former head of public policy evaluation and research projects on participation at the scale of a large watershed (river basin agency), and the person in charge of the territorial animation of water policy at the Ministry of Ecological Transition at the national level. They discuss current trends, key events, main obstacles and levers, as well as anecdotes and recommendations for citizen participation in water management.

The first part of the book addresses the foundations of public participation for socio-ecological sustainability: developing a culture of participation, the profession of territorial facilitator, the construction of social acceptability, the posture of researchers accompanying participatory processes and issues and challenges of e-participation. Chapter 4 is an interview with the Head of the Culture of Public Participation Unit at the General Commission for Sustainable Development at the French Ministry of Ecological Transition and Solidarity. She explains the concept, the objective and functioning of participation charters, which set the values and principles to which the various actors commit and guide the implementation of participatory processes. She details the role of warrants who ensure the sincerity and smooth running of participation. She also evokes the levers for upscaling a culture of participation, namely education, training, more interactions among researchers and policy-makers, reference frameworks and spirits and attitudes. Chapter 5 evokes the profession of territorial facilitators, who support and facilitate participatory processes for the development and conservation of agricultural land in Tunisia. The chapter is based on the testimonies of two facilitators. It shows how this profession seeks to create a link with the population and with all the stakeholders involved. Chapter 6 explores the notion of social acceptability. It argues against participatory approaches that aim to gain acceptance for pre-established technical measures and shows how, in two cases, these approaches have instead opened up a space in which various technical solutions could be discussed. The two cases concern water reuse and artificial wetland buffer zones. Chapter 7 highlights how researchers accompanying participatory processes in support of water policies regularly change their posture, from participation engineers to knowledge transcribers, through trainers of facilitators, evaluators of the participatory process, etc. The chapter includes four testimonies of researchers having adopted different postures in the course of a participatory process. Chapter 8 tackles the issues and challenges when designing a digital platform for supporting participatory policy making. It elicits the potential use conditions and the features provided by various platforms.

The second part of the book addresses altogether the evaluation and engineering of participatory decision-making processes. Chapter 9 focuses on the engineering of participation, i.e. thinking about the objectives, design, choice of methods, implementation, and monitoring and evaluation of a participatory process. The authors present the PrePar tool and identify four key ideas to keep in mind and six structuring questions to ask, to support project leaders in preparing their participatory process. Chapter 10 explains how to evaluate a participatory process: how to assess the participants' demographics while preserving anonymity, how to assess whether all participants could express their opinions, or else how to assess impacts of the process on participants' knowledge, relationships or practices. It discusses issues of task sharing and subjectivity. This chapter includes an insert about the "Participation compass", an app to organise and track participatory processes. Chapter 11 introduces a conceptual framework for assessing the learning effects of participatory processes. It is centered around four main questions: Who learns? What is learned? How does learning take place? And what is learning for? The chapter highlights the need to detail the methodology used to assess learning (when to assess? How to assess? Who assesses? Why assess?) and the contextual and procedural factors impacting learning. The framework is then applied to five case studies in France.

The third part of the book focuses on an approach allowing stakeholders to unveil and make collective decisions about socio-ecological systems in a sustainable and autonomous way: participatory modelling and simulation. Participatory modelling consists of constructing, together with different stakeholders, an object (the model) that allows a number of questions to be answered on a real target system (Minsky, 1965). All five chapters focus on a specific object: role-playing games. Simulation (i.e. the fact of using or running the model) is then used to explore different management options and their social and environmental impacts under different scenarios. Chapter 12 deals with the design and use of role-playing games as methods for implementing participatory approaches for socio-ecological sustainability. It addresses various methodological points about this approach in the form of questions and answers, and then presents the kit for designing the participatory role-playing game "Wat-A-Game" (WAG). Following the chapter, an insert provides a concrete example of a game designed with WAG: LittoWAG is a companion game designed to collect the perception of citizens on the management and adaptation of the coastline to risks. Chapter 13 introduces "L'Eau en Têt", a role-playing game designed with the WAG kit, and used for educational purposes in agricultural high schools in France. Chapter 14 presents WasteWAG (wastewater game), a role-playing game and participatory planning tool for individual and collective sanitation systems designed for urban and rural areas of Senegal. The chapter highlights the singularity of the modelling process, modelled over several successive stages, which contributed to the debate of technical knowledge with local stakeholders. Chapter 15 shows how role-playing-games, often triggered by researchers, may altogether constrain the expression of participants' concerns but also have transformative effects over engineers' vision of local knowledge, reusability of the tools for other purposes, and stakeholders' views on the role of Cambodian preks (drainage canals) in the mosaic landscape. Chapter 16 presents the issues and functioning of participatory observatories. Three examples of participatory observatories of varying duration (from one week to several years) illustrate the diversity of existing observatories and highlight the key role of stakeholders in these mechanisms.

The fourth part of the book presents various tools and processes aiming at co-producing plans toward socio-ecological sustainability. In these experiences, the process is at least as important as the results (i.e. the plans). Chapter 17 presents the CoOPlan approach, aimed at enabling a group of participants to co-construct together a collective strategy (instantiated in an action plan) to change together in their environment. The chapter provides a comparative discussion of the implementation of the CoOPlan approach in four cases (Uganda, metropolitan France, New-Caledonia, Tunisia), highlighting the adaptations made. At the end of the chapter, an insert provides an overview of the French case: a participatory process which engaged over 340 participants in the Drôme river basin in France in order to prepare the revision of the water development and management plan. Chapter 18 summarises the planning process that was implemented in New Caledonia to produce the Shared Water Policy in 2019. The chapter recapitulates the main steps and tools that were used, and what were the main results and feedbacks of participants and organisers. Chapter 19 presents a participatory process implemented in Benin to support the bricolage of local water management institutions. The particularity of this approach is that it combined various tools, including diagnostic, modelling and simulation (role-playing-game), planning and social justice elicitation tools. The approach as a whole was centered on the notion of ecosystem services, with a desire to hybridise the notions of ecosystem services with local knowledge and know-how and to formalise the commitment of the stakeholders concerned to implement sustainable economic alternatives favorable to ecosystems.

The conclusion presents new participatory tools that were being developed during the writing of this book along with pending issues and ways forward.

These 19 chapters address different themes related to socio-ecological systems: agriculture, diffuse pollution, flooding, territorial development, education, sanitation, wetlands, ecosystem services, etc. They also highlight different participatory tools allowing transformations towards socio-ecological sustainability: evaluation, planning, engineering, role-playing-games, observatories, facilitation, etc. Finally, they include cases and examples from eight countries (figure 0.1).

Figure 0.1. Localisation of the cases included in the book. The numbers in the black boxes correspond to the chapters dealing with the cases.

The editors of this book have sought to apply the principles they advocate: participatory, inclusive, transparent and open writing. The book was co-written by 50 researchers and 29 practitioners (decision-makers, politicians, associative actors, territory managers, etc.). It brings together authors and examples from different parts of the world (Figure 0.1). Most of the chapters were written by interdisciplinary teams (management sciences, modelling, agronomy, geography, sociology, economics, etc.) or a-disciplinary teams. The publication is open access which was a sine qua non condition in our choice of publisher. Even the choice of the title (in French) was discussed with all the authors!

▸▸ Acknowledgements

Over more than 20 years, the development of the CoOPLAGE group, methods and tools has been supported by many contributions of researchers, practitioners, interns, administrative staff, who all played a role in this cooperative process. Let them all be acknowledged for their help and inputs (see the list of contributors at the end of the book).

▸▸ References

Arnstein S. R., 1969. A Ladder Of Citizen Participation. *Journal of the American Institute of Planners*, 35(4), 216-224.

Castoriadis C., 1975. *L'Institution imaginaire de la société*, Paris, Seuil.

Glicken J., 2000. Getting stakeholder participation "right": a discussion of participatory processes and possible pitfalls. *Environmental Science & Policy*, 3(6), 305-310.

Luneau A., 2013. Engagement. *In* Casillo I., Barbier R., Blondiaux L., Chateauraynaud F., Fourniau J.-M., Lefebvre R., *et al.* (eds.), *Dictionnaire critique et interdisciplinaire de la Participation, DicoPart* (1ère édition). GIS Démocratie et Participation. https://www.dicopart.fr/engagement-2013

Minsky M., 1965. Matter, Mind and Models. *Proc. of the IFIP Congress*, 45-49, New York.

Chapter 1

Participatory approaches to developing sustainable futures: A global perspective

Katherine Anne Daniell

> This chapter provides a brief overview of the use of participatory processes for developing sustainable futures around the world, with a particular focus on the emergence of participatory methods in the late 1960s and early 1970s. It also reflects on the diversity of current participatory methods. The influences and perspectives of CoOPLAGE in the light of the global context are reflected on, including how its underlying methods stem from a cybernetic, complex systems and engaged political approach. The chapter concludes with potential evolutions and innovations of CoOPLAGE, such as opportunities for integration of emerging technologies and more creative envisioning methods.

⇥ Sustainable futures, pasts and presents?

Involvement of people in developing sustainable futures for their communities is as old as humanity. A diversity of environments across the planet created different needs and interests for societies to manage their survival in relation to these places. Exploitation—without sufficient care and attention to processes of renewal—has led to destruction and death of both humans and the ecosystems sustaining them. This is still the case today, and the balance and process of renewing systems to well-functioning and flourishing states, particularly at the now greatly interconnected global scale, is increasingly fragile.

Where society persists and works with and sculpts their environments through the application of tools and technologies for mutual benefits, ongoing thriving in the same places is made possible. Over time, sometimes over millennia, each one of these social-ecological systems has created specific governance and self-organising systems with rights, responsibilities and relationships to carefully uphold. From the approaches of Caring for Country of Australia's Aboriginal and Torres Strait Islander peoples, including the participatory maintenance of vegetation through mosaic burning and river and aquaculture systems through sophisticated governing arrangements between families and nations like at Baiame's Ngunnhu (the Brewarrina Aboriginal fish traps) (Pascoe, 2014; DCCEEW, 2021) to the terraced agricultural landscapes in South-East Asia, to the rain-farming systems in Africa and oasis management in the Middle East (e.g. Aubriot, 2022), or the Dutch Water boards for managing land and water through it below sea level after having built canal and dyke systems (Dolfing and Snellen, 1999), organised involvement of communities and governing systems to promote long-term maintenance and sustainability is key in their effective functioning (see for example Ostrom 1990; Ostrom *et al.*, 1999; Dietz *et al.*, 2003).

Much of the challenge in many of our current day systems around the planet relates to paces of social-ecological transformation, higher populations of humans, greater and faster engineering of environments, competition for the basics (and not so basic needs) for human and ecosystem thriving, complexity and number of governing (and governing influencing) entities working under a diversity of rules and purposes, which can lead to exclusion, inequity, waste and destruction of the systems on which we all rely. Throughout history such challenges have led to social uprisings and moments of clarity on our collective humanity and how change in governance systems may be necessary to include people usually not making decisions about our collective future.

▸▸ Moments when the potential or challenges for sustainable futures comes into focus: the need for participation

1968 and the beginning of the decade that followed was one of those moments. From the Apollo mission's Earthrise photo and the growing "global" and environmental consciousness[1], to the global student protests and riots including May 68 protests in France (Morin *et al.*, 1968), and one of the first computer art and interactive technology exhibitions in London Cybernetic Serendipity (Reichardt, 1968) it was a period of awakening and developing new processes of participation and engagement that have shaped subsequent generations of practice and research. The research building from this moment included Shelley Arnstein's paper on the Ladder of Public Participation highlighting need for real sharing and moving power of decision-making to citizen control as a part of community organising and urban development in the United States (Arnstein, 1969), the development of the French Groupe des Dix and their cybernetic approaches to rethinking the relations between humans, natures and technologies and how that complexity is better governed by bringing science and politics together (Chamak, 1997; Vivien and Dicks, 2019), to the development of systems dynamics models in North America (Forrester, 1968), new South American pedagogies to overcome oppression (Freire, 1968), and pushes in many parts of the world for Indigenous rights, including the first legal land rights cases in Australia (De Costa, 2006).

In the renewal of democratic thinking and a search for justice of the more marginalised in society, publications like Rawls' (1971) book had a large impact in terms of advocating for larger diversity of views engaged in civil action which will enable greater societal freedom and justice, and Habermas' books (1972, 1984) over the period also led to reflections on legitimation, knowledge and communicative action. Participatory planning and purposeful systems also lay the foundations for the "Search Conferences" of Merriyn and Fred Emery (1974) in Australia. It was also the time of the United Nations Scientific Conference "The Earth Summit" (1972) that set out principles for preserving and enhancing the human environment through international environmental actions. Specifically, this was a period of understanding that many of the traditional operational research methods and their specific quantification principles

---

1. Earthrise, photo taken on December 24, 1968, by Apollo 8 astronaut William Anders, https://www.hq.nasa.gov/office/pao/History/alsj/a410/AS8-14-2383HR.jpg

were not as applicable to social and environmental challenges and research leading to the development of scientific paradigm shifts (Kuhn, 1962) and the birth of the "soft" operational research community where new definitions of these situations included "messes" (Ackoff, 1979), "practical problems" (Ravetz, 1971), "ill-structured problems" (Simon, 1973) and "wicked problems" (Rittel and Webber, 1973; see Rosenhead and Mingers, 2001 for an overview) and later community operations research movements with their emphasis on understanding and working to reduce marginalisation of those typically excluded from decision-making about their lives (Midgley, 2001).

The turn to participatory practice and methods for altering power structures and dominant cultures was also strong in the arts and cultural domains. This included the development and global transmission of forms of emancipatory theatre like Boal's *Theatre of the Oppressed* (1973), which first took hold in South America, where spectators were no longer passive but became "spect-actors" and could through their own action change the direction of the theatre to explore pressing issues. Brand's *Whole Earth Catalog* (1968), published intermittently likewise aimed to give everyone the tools, including ways of thinking in whole systems, methods for self-sufficiency and collective learning, and was important for inspiring counter-culture bottom-up community environmental movements.

However, at this moment, not everything was about concerns of democracy and governance. Researchers were also interested in how to replicate complex systems and evolutionary processes using mathematics and digital computing, and the beginning of research on cellular automata from von Neumann in the early 60s (1966), and many more in the years after (e.g. Arbib, 1966; Yamada and Amoroso, 1969), plus the growth in the use of systems dynamics used in urban systems (Forrester, 1969) and to represent the World's processes through the Club of Rome's publication *The Limits to Growth* (Meadows *et al.*, 1972), set the scene for simulation modelling and games to understand interacting behaviours, whether human, biological, ecological or strategic across space and time. These were vital in the development of models and representations of what systems and their states through time could be considered to be sustainable and/or resilient, and what might need to change to navigate them in such directions. Bringing all these areas together, changing societies, technologies and environments, it is not surprising it was also one of the core moments when traditional scientific practice was challenged and its ill fit to globally interconnected challenges outlined across many disciplines leading to the development of transdisciplinary and participatory research and praxis (Lassudrie-Duchêne, 1968; Piaget, 1972; Jantsch, 1972). Echos of the challenges of this period can be found both through many centuries and decades of history and in the following years, as all these (r)evolutions built on and lay the foundations for other participatory practices and managing the challenges of sustainability through other moments of change and awareness of the need for alternative approaches to navigating complex systems.

▸▸ A diversity of participatory practices

Fast forward through the decades, and the diversity of approaches to participatory practice for navigating towards more sustainable futures continues to grow and evolve across the world. The development of computational and communications infrastructures has enabled a range of new systems for gathering and structuring diverse inputs from a range of sources (e.g. sensing technologies, published content) and people interested in exploring systems and making decisions on future individual and collective actions in relation to them. The period from the late 1990s to late 2000s was a period of particular period of growth and development, as limits of natural resource management systems and actions to kerb climate change (e.g. the Kyoto Protocol in 1997), under often what appeared to be technocratic regimes in democracies, again came into stark view.

Common families of participatory approaches which are often used in conjunction with each other include:
– Voting and preference gathering systems, including for participatory decision support (e.g. Rios Insua *et al.*, 2008)
– Traditional public meeting structures (e.g. Field, 2019)
– Participatory and group model building and role-playing/simulation communities (e.g. Voinov and Bousquet, 2010; Abrami *et al.*, 2021)
– Yarning and story-based exchange (e.g. Yunkaporta and Kirby, 2011)
– Participatory and collaborative design (e.g. Negroponte, 1975)
– Deliberative democracy methods including mini publics and citizen juries/ consensus conferences (e.g. Gastil and Lavine, 2005; Dryzek *et al.*, 2019)
– Participatory theatre and creative public engagement including participatory photography and creative writing/musical improv. (e.g. Conrad and Sinner, 2015)
– Immersive cybernetic art and installations (e.g. Pickering, 2024; Jacucci *et al.*, 2010)
– Participatory mapping and planning, including participatory rural appraisal and participatory geographic information system (GIS, e.g. Cochrane and Corbett, 2020)
– Futuring and prospective methods including scenario methods, science fiction writing/prototyping (e.g. Wyborn *et al.*, 2021; Bishop, 2011; Johnson, 2011; Alexandra *et al.*, 2023)
– Open source communities, citizen science and participatory evaluation (Conrad and Hilchey, 2011; Cullen and Coryn, 2011; Eghbal, 2020)
– Social media and Information and communications technology (ICT)-supported participation (e.g. Lin and Kant, 2021)
– Multi-level and collaborative governance, including group decision support (Huxham, 1996; Bache and Flinders, 2004; Daniell and Kay, 2017)
– Conflict mediation/transformation including dispute resolution, negotiation and restorative justice (e.g. Delli Priscoli, 2003; Susskind *et al.*, 1999)
– Collaborative engineering of participatory processes (e.g. Kolfschoten *et al.*, 2006; Daniell, 2012; Ferrand *et al.*, 2021).

Each family often has an underlying politic and set of assumptions/purposes on knowledge processes, their interactions and the futures and impacts they envisage. How they seek to influence action in the world, including policy processes, is typically linked to the positionality of the convenors of the processes (Daniell *et al.*, 2016a), including cultural, disciplinary and political orientations, and how they seek to communicate and govern with, or over, others (Follett, 1924; Fung, 2006). Each new participatory process descended from, and carries a mix of elements from, these families and more. It then creates a unique set of changes, new relationships and knowledge about the world and imaginaries for its future; see for example Nabavi (2022) on blending improv' theatre and futuring for developing transformative engagement on water conflict.

Many common orientations in participatory approaches to sustainability, particularly the development of new approaches, seek challenge or blur the boundaries of the roles of "experts", "citizens" and "decision makers", and to adjust existing power structures and regimes of expertise/knowledge (Thomas, 2004). Through this process, more views are included and this multiplicity negotiated through the use of specific processes and "intermediary objects" or technologies. These can take the form of maps or models/representations through any media, which instead moves power struggles and the specific types of expertise needed to those who can effectively design and facilitate/implement processes employing these specific modalities (Daniell *et al.*, 2010) and induces pressure to do it well or risk disempowered and disappointed stakeholders (Barreteau *et al.*, 2010). This has led to the development of more participatory processes at the level of the design and implementation of these processes, including "self-designed" participatory modelling and simulation processes (e.g. D'Aquino *et al.*, 2003) and full participatory process structures and toolkits like CoOPLAGE (e.g. Ferrand *et al.*, 2021).

⇥ CoOPLAGE as a participatory toolkit in the global context: influences and perspectives

CoOPLAGE and the underlying and associated approaches outlined in this book draw on many strands of the rich theory and practical approaches to participation above. In particular, CoOPLAGE and the underlying methods take a cybernetic, complex systems and engaged political approach (Ferrand *et al.*, 2021), with what could be identified as a specifically French interpretation that reflect on power, politics and knowledges in an explicit and transparent way. It is particularly cybernetic in that it focusses on the "participatory process of/for the participatory process" including the first tools of the kit PrePar and CreaWAG that seek to support groups of people to design their own participatory processes and models. In addition, CoOPLAGE has a set of embedded values, beliefs and a politic which is an orientation to intervention research (David, 2000; Midgley, 2001), and on-the-ground decision and planning support. Specifically, the toolkit and the researchers behind it consider that research has a positive role to play in society, and that the acts of researchers can change the direction of societal transformations and provide structured and open spaces and methods for participation, knowledge sharing and construction of those typically making, and being affected by, decisions, together. CoOPLAGE does typically not seek to create citizen deliberation in search of consensus separately from those with decision-making power, or work on representative principles, as in some deliberative democracy instances (see for example Fishkin and Mansbridge, 2017). It rather works on a process of organising knowledges, perspectives and values at multiple levels in an Ostromian sense: at the management action level or the arena of operational choice, as well as the arenas of collective and constitutional choices (Ostrom, 1990). CoOPLAGE is intended to give power to groups of people and communities to organise their own knowledge and struggles to govern the commons together, and to gain rapid feedback through a range of simulations and evaluation processes on the potential for their individual and collective actions to create change (Daniell, 2012; Hassenforder *et al.*, 2019, 2021).

The main institutional group involved in the development and use of CoOPLAGE and its underlying methods is the G-EAU joint research unit "Water Matters". This group has been strongly influenced and shaped through praxis by the French education system and its orientation at higher degree level to applied public service, including internationally, particularly through the generalist "Grandes Écoles" and national "Grands Corps de l'État" (state public service corps personnel), specifically the rural, water and forestry engineers[2] or bridges, water and forestry engineers[3] (Igref or Ipef since 2009). All of these institutions and the groups of people within them have a long history of mixing disciplinary research backgrounds around social-ecological policy theme, development situation or system of interest (see for example the papers in the French journal *Natures, Sciences, Sociétés*). Moreover, the interdisciplinary systems, cybernetics and management sciences, socio-political sciences and environmentally-aware engineering approaches, which are deeply embedded in these education and orientations of public service research systems, have provided fertile ground for collaborative and transdisciplinary development of collective engineering approaches to participatory processes for social-ecological systems management and for governing in regional and multi-level regional approaches. Within the CoOPLAGE community there is still a diversity of approaches depending upon choices and involvement on different types of modelling, simulation and role-playing games (some mediated through computer simulation models—often multi-agent modelling, originally through the COmmon pool Ressources and Multi-Agent Simulations platform – Cormas, e.g. Bousquet *et al.*, 1998) but often low-tech using local or imported materials—stones, cards, cups to represent parts of the environment and circulating flows of water and economic and system production changes (e.g. fish stocks, agricultural production, wetland quality—see for example Wat-A-Game, Abrami *et al.*, 2016). There is also a willingness to co-test methods and train local and international facilitators to help with implementing and evaluating methods (Hassenforder *et al.*, 2016). CoOPLAGE represents, however, a relatively specific set of tools and methods that is of Western axiological, ontological, epistemological and methodological origins. The approach is one of inclusive engineering, although interventionist in time and space that is construed in Western ways, even if there is some space for holding other belief and value systems through the process. CoOPLAGE processes can sometimes present as structurally violent compared to other cultures of participation, interaction and communication such as yarning circles, arts and creative storytelling-based methods, and even some types of discursive deliberation focused efforts. That said, it has also been accepted—taking into account the creative cybernetic leadership principal of "productive discomfort" (Gould *et al.*, 2022)—by communities seeking two-ways (Country *et al.*, 2015; RiverOfLife *et al.*, 2021), or multi-ways governance, and who are willing to come together for social learning and collectively developing plans for the future (e.g. Lejars *et al.*, 2021; Daniell *et al.*, 2016b).

Within the CoOPLAGE approach there is indeed potentially space for joining multiple traditions of participatory practice together (e.g. participatory theatre, specific forms of mediated deliberation on certain decision objects, artistic representations of systems in addition to the set of regularly employed frameworks and tools), although the particular politic drives more towards the creation of distributed "actionable knowledge", where all participants can be actors for change and have the ability to coordinate these for "effective" individual and collective action, defined on their own terms and in line with their own value and belief systems.

---

2. Ingénieurs du génie rural, des eaux et des forêts (Igref)

3. Ingénieurs des ponts, des eaux et des forêts (Ipef)

The CoOPLAGE toolkit encourages mapping out different perspectives then rapidly coming to a common point of view and understanding (through shared collective artefacts these views are recorded on) on this diversity and collective ways forward. Although the approach can include methods of speculation or *prospective*, CoOPLAGE and the research teams supporting its use are not necessarily interested in defining and/or articulating *competing* worldviews and politics as it then makes choice and negotiation trickier. Rather the approach aims to seek acknowledgment of a collection of viewpoints, and a set of potential options/futures, as part of a whole system, rather than setting up "teams" like in a debate. It therefore is not focused on groups winning or losing, but seeks its ideal of just and equitable sharing of common resources (Rawls, 1971; Neal *et al.*, 2014) and moving forward collectively. This participatory mode of collective action planning can potentially clash with many Western governance representational democracy settings, which are often set up in a majority and competing ideologies mode. This means that for CoOPLAGE who gets to choose who will be in the room, and those people's relationships to decision makers, are particularly important to ensure capacity for action. It also means convenors need to build trust with key actors in the systems and to get them on board with using the toolkit methods for their systems.

►► Potential CoOPLAGE evolutions and innovations to support sustainable futures

How CoOPLAGE might interface with emerging technologies and other opportunities for participatory processes is also worth reflecting on, and indeed to some extent has already been discussed in the community (e.g. see Rios Insua and French (2010) for discussions and methods of eDemocracy). In terms of advanced modelling and analytical techniques, including real-time monitoring and artificial intelligence (AI)-based alert systems that could support participatory processes, there have already been attempts to incorporate such systems and knowledge within examples of the community. This is particularly the case when CoOPLAGE type methods could be coupled with territorial intelligence systems—see examples in Daniell *et al.* (2020) such as those in the Herault, e.g. Ouest Herault (Dionnet and Guérin-Schneider, 2014), Bassin de Thau, regions of France where charters of participation, participatory process design, role-playing games and remote-sensing territorial approaches are being used in close proximity; and in the Lower Hawkesbury Estuary where participation design, participatory modelling and real-time AI-based water quality monitoring, particularly for algal blooms co-exist (Coad *et al.*, 2014)—and can be drawn on as both knowledge inputs to processes and ongoing monitoring and evaluation systems of environmental conditions. Other types of analytical techniques may underlie online participatory systems and depending on the purposes and techniques may be compatible with the CoOPLAGE politic and methodology, although to what extent the common artefacts can be built and trusted, may depend greatly on platform and/or facilitator capabilities to make sense of the online systems and common artefacts to participants. The face-to-face domain has to date been where these common artefact-based systems have thrived, as CoOPLAGE and its internal methods seek not only to create a series of common artefacts and collective templates for action, but through the social processes around these, build trust and understanding between participants linked to a common experience and orientation of responsibility for collective action. To what extent online systems can be partly or wholly used for this—first those already part of the toolkit (Ferrand *et al.*, 2021) and then others like engaging with metaverse-type systems and online immersive augmented/virtual reality (AR/VR) gaming platforms for collective modelling, envisioning and planning (e.g. Evans *et al.*, 2022; Hudson-Smith and Shakeri, 2022)—remain research questions not just for the CoOPLAGE community but other participatory researchers around the world.

Likewise, development of scenarios and creative envisioning and prospective methods as elements of participatory practice have been evolving in recent years, both in their creativity but also in their philosophical and mathematical bases (e.g. Lord *et al.*, 2016; Dourish and Bell, 2011; Conrad and Sinner, 2015; Bell, 2021 in the North American and Australian traditions). There are opportunities for these to be brought into or interfaced with the CoOPLAGE methods, as long as a strong enough collective approach to a way forward can be fostered.

In addition, innovation in voting systems and online deliberative "liquid" or "crypto" democracy techniques (e.g. Engin, 2016; Allen *et al.*, 2020) may be able to feed into CoOPLAGE-mediated processes, but with quite different underlying philosophical approaches, they are more likely to be used by different communities for different reasons.

There remain opportunities for learning from participatory approaches with similar politics in other domains such as health care, and educational space design, although the strong Ostromian backbone and orientation to common pool and particularly natural resources may work against some elements of easy translation to completely different systems in the short term. CoOPLAGE and the examples in this book are highly relevant, and now globally tested, in many countries and sustainability contexts. The orientation of CoOPLAGE to action and empowerment of community members deploying the methods themselves to plan their own participatory processes for their own futures is one that, to my knowledge, is unique in the global context, and full of potential for supporting greater numbers of communities to co-create their own just and sustainable futures.

⮞ References

Abrami G., Daré W.S., Ducrot R., Salliou N., Bommel P., 2021. Participatory modelling. *In: The Routledge Handbook of Research Methods for Social-Ecological Systems*, Routledge, 189-204.

Abrami G., Ferrand N., Ducrot R., Morardet S., Hassenforder E., Garin P., *et al.*, 2016. Paper and pebbles simulations and modelling for the governance of socio-environmental systems: a review of 8 years of experimenting with the Wat-A-Game toolkit. *In*: Sauvage S., Sánchez-Pérez J.M., Rizzoli A.E. (eds), *Proceedings of the 8th International Congress on Environmental Modelling and Software* (iEMSs 2016). 10-14 July 2016, Toulouse, France, 851, 5 vols (1335 p.)

Ackoff R.L., 1979. The future of operational research is past, *Journal of the Operational Research Society*, 30, 93-104.

Alexandra C., Wyborn C., Roldan C.M., van Kerkhoff L., 2023. Futures thinking: concepts, methods and capacities for adaptive governance. *In*: Juhola S. (ed.) *Handbook of Adaptive Governance*, Edward Elgar Publishing, 76-98.

Allen D.W., Berg C., Lane A.M., Potts J., 2020. Cryptodemocracy and its institutional possibilities. *The Review of Austrian Economics*, 33, 363-374.

Arbib M., 1966. Simple self-reproducing universal automata, *Information Control*, 9, 177-189.

Arnstein S.R., 1969. A ladder of citizen participation, *Journal of the American Institute of planners*, 35(4), 216-224.

Aubriot O., 2022. The history and politics of communal irrigation: A review, *Water alternatives*, 15(2), 307-340.

Bache I., Flinders M. (eds), 2004. *Multi-level governance*. Oxford University Press, Oxford.

Barreteau O., Bots P.W., Daniell, K.A., 2010. A framework for clarifying "participation" in participatory research to prevent its rejection for the wrong reasons. *Ecology and Society*, 15(2). http://www.ecology-andsociety.org/vol15/iss2/art1/

Bell G., 2021. Touching the future: Stories of systems, serendipity and grace. *Griffith Review*, 71, 251-262.

Bishop P., 2011. Scenario Development, Epistemology and Methods. *In:* Nováček P., Schauer T. (eds) *Learning from the Futures*, Olomouc, Palacký University Press, 11-23.

Boal A., 1973. *Theatre of the Oppressed*, Routledge.

Bousquet F., Bakam I., Proton H., LePage C., 1998. Cormas: CommonPool Resources and Multi-Agent Systems. *Lecture Notes in Artificial Intelligence*, 1416, 826–837.

Brand S. (ed.), 1968. *The Whole Earth Catalog: Access To Tools*, Portola Institute.

Chamak B., 1997. *Le Groupe des Dix ou les avatars des rapports entre science et politique*, Paris, Éditions du Rocher.

Coad P., Cathers B., Ball J.E., Kadluczka, R., 2014. Proactive management of estuarine algal blooms using an automated monitoring buoy coupled with an artificial neural network. *Environmental modelling & software*, 61, 393-409.

Cochrane L., Corbett J., 2020. Participatory mapping. *In*: *Handbook of communication for development and social change*, pp.705-713.

Conrad C.C., Hilchey K.G., 2011. A review of citizen science and community-based environmental monitoring: issues and opportunities. *Environmental monitoring and assessment*, 176, 273-291.

Conrad D., Sinner A. (eds), 2015. *Creating together: Participatory, community-based, and collaborative arts practices and scholarship across Canada*. Wilfrid Laurier University Press.

Country B., Wright S., Suchet-Pearson S., Lloyd K., Burarrwanga L., Ganambarr R., *et al.*, 2015. Working with and Learning from Country: Decentring Human Author-Ity, *Cultural Geographies*, 22(2), 269-83.

Cullen A., Coryn C.L., 2011. Forms and functions of participatory evaluation in international development: A review of the empirical and theoretical literature. *Journal of MultiDisciplinary Evaluation*, 7(16), 32-47.

Daniell K.A., White I., Ferrand N., Ribarova I.S., Coad P., Rougier J.E., *et al.*, 2010. Co-engineering participatory water management processes: theory and insights from Australian and Bulgarian interventions. *Ecology and Society*, 15(4). https://www.ecologyandsociety.org/vol15/iss4/art11/

Daniell K.A., 2012. *Co-engineering and participatory water management: organisational challenges for water governance*, Cambridge University Press, Cambridge, UK.

Daniell K.A., Morton A., Ríos Insua D., 2016a. Policy analysis and policy analytics. *Annals of Operations Research*, 236, pp.1-13.

Daniell K.A., Bouard S., Poelina A., Ferrand N., White I., Alexandra J., *et al.*, 2016b. *Investigating the Challenges of Water Governance in Oceania*, Water, Infrastructure and the Environment Conference, 28 Nov-2 Dec, Queenstown, New Zealand.

Daniell K.A., Kay A., 2017. *Multi-level governance: conceptual challenges and case studies from Australia.* ANU Press, Canberra, 474p.

Daniell K.A., Plant R., Pilbeam V., Sabinot C., Paget N., Astles K., *et al.*, 2020. Evolutions in estuary governance? Reflections and lessons from Australia, France and New Caledonia. *Marine Policy*, 112, 103704. https://www.sciencedirect.com/science/article/pii/S0308597X18303336

D'Aquino P., Le Page C., Bousquet F., Bah A., 2003. Using self-designed role-playing games and a multi-agent system to empower a local decision-making process for land use management: The SelfCormas experiment in Senegal. *Journal of artificial societies and social simulation*, 6(3). https://ideas.repec.org/a/jas/jasssj/2003-19-2.html

David A., 2000. La recherche-intervention, cadre général pour la recherche en sciences de gestion ? *In*: David A., Hatchuel A., Laufer R. (ed.), *Les nouvelles fondations des sciences en gestion*, Paris, Vuibert, collection FNEGE, 193-213.

Dionnet M., Guérin-Schneider L., 2014. La coordination inter-organisationnelle, levier de la gouvernance territoriale : quelles leçons tirer de la gestion de l'eau interbassin ? *Géographie Économie Société*, 16(4), 399-420.

DCCEEW, 2021. *National Heritage Places – Brewarrina Aboriginal Fish Traps (Baiame's Ngunnhu)*, Department of Climate Change, Energy, the Environment and Water, Australian Government. https://www.dcceew.gov.au/parks-heritage/heritage/places/national/brewarrina

De Costa R., 2006. Identity, authority, and the moral worlds of indigenous petitions, *Comparative Studies in Society and History*, 48(3), 669-698.

Delli Priscoli J., 2003. *Participation, Consensus Building, and Conflict Management Training Course (Tools for Achieving PCCP)*. IHP-IV Technical documents in Hydrology PC->CP Series, No 22. Paris, France: UNESCO-IHP prepared for the WWAP.

Dietz T., Ostrom E., Stern P.C., 2003. The struggle to govern the commons, *Science*, 302(5652), 1907-1912.

Dolfing B., Snellen W.B., 1999. *Sustainability of Dutch water boards: Appropriate design characteristics for self-governing water management organisations*, ILRI, Wageningen, 52p. https://library.wur.nl/WebQuery/wurpubs/fulltext/78801

Dourish P., Bell G., 2011. *Divining a digital future: Mess and mythology in ubiquitous computing*, Cambridge, MA, MIT Press.

Dryzek J.S., Bächtiger A., Chambers S., Cohen J., Druckman J.N., Felicetti A. *et al.*, 2019. The Crisis of Democracy and the Science of Deliberation. *Science* 363(6432), 1144-46. https://doi.org/10.1126/science.aaw2694

Eghbal N., 2020. *Working in public: the making and maintenance of open source software*, San Francisco, CA, Stripe Press.

Emery F.E., Emery M., 1974. *Participative design: Work and community life.* Centre for Continuing Education, Canberra, Australian National University.

Engin T., 2016. Can a new app save democracy? Masters of Media, University of Amsterdam, http://mastersofmedia.hum.uva.nl/blog/2016/09/19/can-a-new-app-save-democracy/

Evans L., Frith J., Saker, M., 2022. *From Microverse to Metaverse: Modelling the Future through Today's Virtual Worlds*. Bingley, UK, Emerald Group Publishing.

Ferrand N., Hassenforder E., Aquae-Gaudi, W., 2021. Quand les acteurs modélisent ensemble leur situation, principes ou plans pour décider et changer durablement, en autonomie, *Sciences Eaux & Territoires pour tous*, 35, 14-23.

Field J.B., 2019. *Town Hall Meetings and the Death of Deliberation.* University of Minnesota Press.

Fishkin J.S., Mansbridge J. (eds), 2017. The prospects & limits of deliberative democracy. *Dædalus: Journal of the American Academy of Arts & Sciences*, Summer 2017, 171p.

Follett M.P., 1924. *Creative experience.* Longmans, Green and company.

Forrester J.W., 1968. *Principles of Systems* (2nd ed.), Waltham, MA, Pegasus Communications, 391p.

Forrester J.W., 1969. *Urban Dynamics.* Cambridge, Mass.: M.I.T. Press.

Freire P., 1968. *Pedagogy of the oppressed*, New York, The Continuum Publishing Company.

Fung A., 2006. Varieties of Participation in Complex Governance. *Public Administration Review*, 66(s1), 66–75.

Gastil J., Levine P., (eds), 2005. *The Deliberative Democracy Handbook: Strategies for Effective Civic Engagement in the Twenty-First Century*. San Francisco: Jossey-Bass.

Gould M., Daniell K.A., Meares A., Bell G., 2022. *Re/defining Leadership in the 21st century: the view from cybernetics*, Whitepaper, Australian National University & Menzies Foundation, Canberra, Australia. https://cybernetics.anu.edu.au/assets/Redefining Leadership in the_21st_Century-the_view_from_Cybernetics.pdf

Habermas J., 1972. *Knowledge and Human Interests*. New York, Beacon Press.

Habermas J., 1984. *The Theory of Communicative Action*. Vols I and II, 3rd edn. Oxford, UK, Polity Press.

Hassenforder E., Ferrand N., Girard S., Barreteau O., 2019. *Citizen Science for Public Decision Making and Social Change: when scientific objectivity is challenged*. International workshop "Citizen science: new epistemological, ethical and political challenges", Jun 2019, Lyon, France. https://hal.inrae.fr/hal-02609642

Hassenforder E., Girard S., Ferrand N., Petitjean C., Fermond C., 2021. La co-ingénierie de la participation : une expérience citoyenne sur la rivière Drôme, *Nature Sciences Sociétés*, 29, 159-173.

Hassenforder E., Pittock J., Barreteau O., Daniell K.A., Ferrand N., 2016. The MEPPP framework: a framework for monitoring and evaluating participatory planning processes. *Environmental management*, 57, 79-96.

Hudson-Smith A., Shakeri M., 2022. The Future's Not What It Used To Be: Urban Wormholes, Simulation, Participation, and Planning in the Metaverse. *Urban Planning*, 7(2), 214-217. https://doi.org/10.17645/up.v7i2.5893

Huxham C. (ed.), 1996. *Creating collaborative advantage*, London, UK, Sage Publications.

Jacucci G., Wagner M., Wagner I., Giaccardi E., Annunziato M., Breyer N., *et al.*, 2010. *ParticipArt: Exploring participation in interactive art installations*. 2010 IEEE International Symposium on Mixed and Augmented Reality-Arts, Media, and Humanities, IEEE, 3-10.

Jantsch E., 1972. Towards interdisciplinarity and transdisciplinarity in education and innovation. In, Centre for Educational Research and Innovation (CERI) (ed.), *Interdisciplinarity: Problems of teaching and research in universities* (pp. 97-121). Paris, France: Organisation for Economic Co-operation and Development.

Johnson B.D., 2011. Science fiction prototyping: Designing the future with science fiction. *Synthesis Lectures on Computer Science*, 3(1), 1-190.

Kolfschoten G.L., Briggs R.O., De Vreede G.J., Jacobs P.H., Appelman J.H., 2006. A conceptual foundation of the thinkLet concept for Collaboration Engineering. *International Journal of Human-Computer Studies*, 64(7), 611-621.

Kuhn T.S., 1962. *The structure of scientific revolutions*, Chicago, University of Chicago Press.

Lassudrie-Duchêne B., 1968. Economie politique et sociologie : rapport au congrès des économistes de langue Française, *Revue d'économie politique*, mai-juin 1968, Vol. 78, No. 3, Domaines et méthodes des sciences économiques, 414-447.

Lejars C., Bouard S., Ferrand, N., 2021. La politique de l'eau partagée en Nouvelle-Calédonie : retour d'expériences sur un dispositif de co-construction et de co-planification. *Sciences Eaux & Territoires*, 35, 68-75.

Lin Y., Kant S., 2021. Using social media for citizen participation: Contexts, empowerment, and inclusion. *Sustainability*, 13(12), 6635.

Lord S., Helfgott A., Vervoort, J.M., 2016. Choosing diverse sets of plausible scenarios in multidimensional exploratory futures techniques, *Futures*, 77, 11-27.

Meadows D.H., Meadows D.L., Randers J., Behrens III, W.W., 1972. *The limits to growth: a report to the club of Rome on the predicament of mankind*, New York, Universe Books.

Midgley G., 2001. *Systemic intervention: Philosophy, methodology, and practice*, Springer Science & Business Media.

Morin E., Lefort C., Castoriadis C., 1968. *Mai 1968 : la Brèche, premières réflexions sur les événements*, Fayard, Paris, 142p.

Nabavi E., 2022. Pathways theatre: Using speculative and collaborative improvisation for transformative engagement. *Futures*, 143, 103022.

Neal M. J., Lukasiewicz A., Syme G. J., 2014. Why justice matters in water governance: some ideas for a "water justice framework." *Water Policy*, 16, 1-18.

Negroponte N., 1975. *Soft architecture machines*, Cambridge, MA, MIT Press.

Ostrom E., 1990. *Governing the commons: the evolution of institutions for collective action*, Cambridge, UK, Cambridge University Press.

Ostrom E., Burger J., Field C.B., Norgaard R.B., Policansky D., 1999. Revisiting the commons: local lessons, global challenges, *Science*, 284(5412), 278-282.

Pascoe B., 2014. *Dark Emu: Aboriginal Australia and the birth of agriculture*, Broome, Magabala Books.

Piaget J., 1972. The epistemology of interdisciplinary relationships. In: Centre for Educational Research and Innovation (CERI) (ed.), *Interdisciplinarity: Problems of teaching and research in universities* Paris, France: Organisation for Economic Co-operation and Development, 127-139.

Pickering A., 2024. Cybernetic art. *In:* Gere C. (ed.) *The encyclopaedia of new media art,* Bloomsbury Publishing.

Rawls J., 1971. *A theory of justice.* Cambridge, MA, Belknap Press.

Ravetz J.R., 1971. *Scientific Knowledge and its Social Problems*, Oxford, UK, Oxford University Press.

Reichardt J. (ed), 1968. *Cybernetic Serendipity: The Computer and the Arts*, Studio International Special Issue, London, July, 101p.

Rios Insua D., Kersten G.E., Rios J., Grima, C., 2008. Towards decision support for participatory democracy. *Information systems and e-business management*, 6, 161-191.

Rios Insua D., French S. (eds.), 2010. *E-democracy: a group decision and negotiation perspective* (Vol. 5). Springer Science & Business Media.

RiverOfLife M., Taylor K.S., Poelina A., 2021. Living waters, law first: Nyikina and Mangala water governance in the Kimberley, Western Australia, *Australasian Journal of Water Resources*, 25(1), 40-56.

Rittel H.J., Webber M.M., 1973. Dilemmas in the general theory of planning, *Policy Science*, 4, 155-169.

Rosenhead J., Mingers J., 2001. *Rational analysis for a problematic world revisited*, Chichester, John Wiley and Sons Ltd.

Simon H.A., 1973. The structure of ill-structured problems, *Artificial Intelligence*, 4, 181-201.

Susskind L., McKearnan S., Thomas-Larmer J. (eds), 1999. *The Consensus Building Handbook: A Comprehensive Guide to Reaching Agreement.* Thousand Oaks, CA, Sage Publications.

Thomas R. L., 2004. Management of freshwater systems: the interactive roles of science, politics and management, and the public. *Lakes and Reservoirs: Research and Management*, 9(1), 65-73.

Vivien F.D., Dicks H., 2019. Dossier : Le groupe des Dix, des précurseurs de l'interdisciplinarité – Introduction– De la légende dorée aux querelles d'héritage : le Groupe des Dix en transversal, *Natures Sciences Sociétés*, 27(2), 125-136.

Voinov A., Bousquet, F., 2010. Modelling with stakeholders. *Environmental modelling & software*, 25(11), 1268-1281.

von Neumann J., 1966. *Theory of Self-Reproducing Automata*, Edited and completed by A.W. Burks, Urbana, University of Illinois Press.

Wyborn C., Louder E., Harfoot M., Hill S., 2021. Engaging with the science and politics of biodiversity futures: a literature review, *Environmental Conservation*, 48(1), 8-15.

Yamada H., Amoroso S., 1969. Tessellation automata, *Information Control.* 14, 299-317.

Yunkaporta T., Kirby M., 2011. Yarning up Indigenous pedagogies: A dialogue about eight Aboriginal ways of learning. *In:* Bell H.R. Milgate G., Purdie N. (eds), *Two Way Teaching and Learning: towards culturally reflective and relevant education*, Camberwell, Victoria, ACER Press.

Chapter 2

The CoOPLAGE approach: When actors model their situation, principles or plans together for sustainable, empowering decision-making and change

Nils Ferrand, Emeline Hassenforder and Wanda Aquae-Gaudi[1]

CoOPLAGE is the acronym for "Coupler des outils ouverts et participatifs pour laisser les acteurs s'adapter pour la gestion de l'environnement[2]. This approach aims at guiding stakeholder participation (citizens, elected officials, managers, etc.) in the decision-making process with regard to their environment. This chapter presents the fundamental principles of the CoOPLAGE approach (empowerment, intervention research, true participation in decision-making, reflexivity on desired changes as well as a mix of engineering and do-it-yourself). In line with works on the modelling of complex systems, the background of this approach is also reviewed here. Lastly, the various CoOPLAGE tools are introduced, then detailed in the different chapters of this book.

CoOPLAGE is a set of complementary tools designed to meet the needs of stakeholders in supporting socio-environmental transition. With these tools, stakeholders can:
– share their views of a socio-environmental situation,
– explore the outcomes of their practices and choices in terms of public policy,
– choose how to organise decision-making and assign roles,
– discuss principles of justice,
– propose action plans to deal with complex issues, and
– monitor and evaluate where they stand in their change process.

The CoOPLAGE suite of tools has been built over the years by researchers from the G-EAU joint research unit "Water Matters" in Montpellier in response to the decision-making needs of their field partners in various operational projects in France and abroad.

With and for all stakeholders, the participatory modelling process is at the heart of the CoOPLAGE approach (Box 2.1). Participatory modelling consists of constructing,

---

1. Wanda Aquae-Gaudi is a fictional author created in 2010 to represent the CoOPLAGE collective. With more than 100 contributors since 2008, it was necessary to recognise the contributions of everyone in the design of methods and scientific productions. Wanda's list can be found at the end of the book.
2. Coupling Open and Participatory Tools to Let Actors Adapt for Environmental Management

Role-play representing a socio-environmental system
(Wat-A-Game tool)

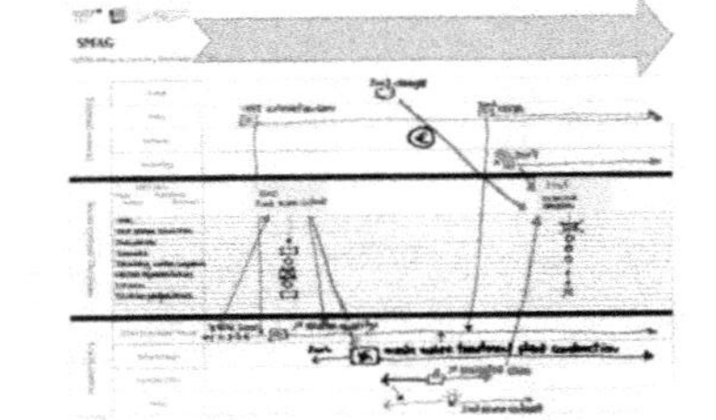

Natural resource management
plan representing a management
strategy (CooPlan tool)

Participation plan representing a decision-making process
(PrePar tool)

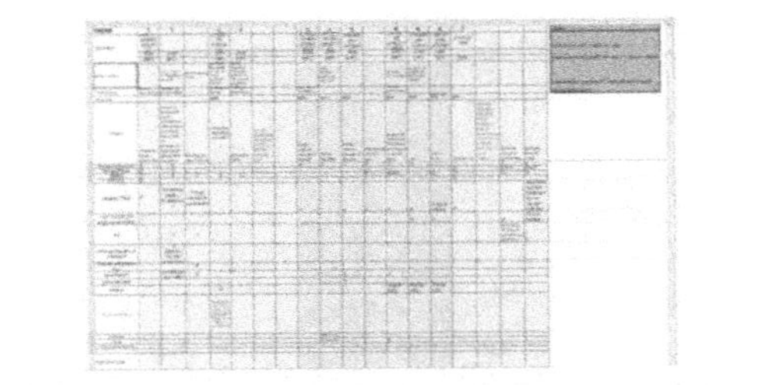

Timeline representing the past governance of a territory
(SMAG or self-modelling for assessing governance tool)

Figure 2.1. Examples of models representing different systems
(tools mentioned in parenthesis are explained below and in the different chapters of this book)

together with different stakeholders, an object (the model) that allows a number of questions to be answered on a real target system (Minsky, 1965). The object in question can be, for example, a role-playing game, a timeline, a map, a diagram or a matrix. The system represented by this object can be, among other things, a territory, a decision-making process or a management strategy (figure 2.1). The idea is that the object, or model, should enable the various actors to step back from the system, so that they may ask themselves the right questions, consider its various components and take a fresh look at it. The object thus acts as a kind of critical mirror of the system to support collaborative decision-making between the actors. But beyond the ultimate use of the object in decision-making, what is important is the construction of the object as such (the modelling). By building a common representation of their system, the stakeholders learn to work together, exchange their different views, and take ownership of the issues and actions to be taken. They thus build the conditions for their own empowerment and collaboration towards socio-environmental transition.

This approach is therefore very different from classic coordination approaches where models, options, choices and regulations are provided by external, technical, administrative or political actors. Even when these approaches are guided by information sharing, consultation or light forms of communicative participation, they are still perceived by those in action as being controlled by experts and decision-makers, and therefore outside their own control and responsibility.

In what follows below and in the various chapters of this book, we will see how the CoOPLAGE approach can be concretely implemented in the field via different principles and methods. The rest of this chapter is devoted to positioning CoOPLAGE as an instrument for supporting socio-environmental transition.

Box 2.1. Historical background behind the CoOPLAGE approach

CoOPLAGE participatory modelling is in line with works on modelling complex industrial or socio-environmental systems that followed and were based on Jay Forrester's (1968) system dynamics and his famous World II model, which backed "The Limits to Growth" (Meadows *et al.*, 1972) and the opinions expressed by the Club of Rome (figure 2.2). Modelling linking society and the environment has been present in France since early precursory works on "cybernetics" that were extended to socio-economic systems (Moles, 1968; Wiener, 1950). Cybernetics is a science that exclusively studies communications and their regulation in natural and artificial systems (Wiener, 2019). It allows for all encountered mechanisms to be explained and understood using a few simple logical building blocks, such as the emitter (which emits information), the receiver (which receives information) and the feedback (action of an effect on its own origin).

However, it is essentially the work on ecological or epidemiological modelling that has led to the questioning of interdisciplinarity and the linking of models, which also required bringing people together (Pave and Jollivet, 1993; Schmidt-Lainé and Pavé, 2002) and, in France, initiating and supporting the cross-cutting environment-life-society programme by key figures (J.-M. Legay, M. Jollivet, A. Pavé, J. Weber, S. Van Der Leeuw).

In the early 1990s, a trend towards complex systems, their modelling and ultimately their control appeared. This trend mobilised, on the one hand, a more theoretical

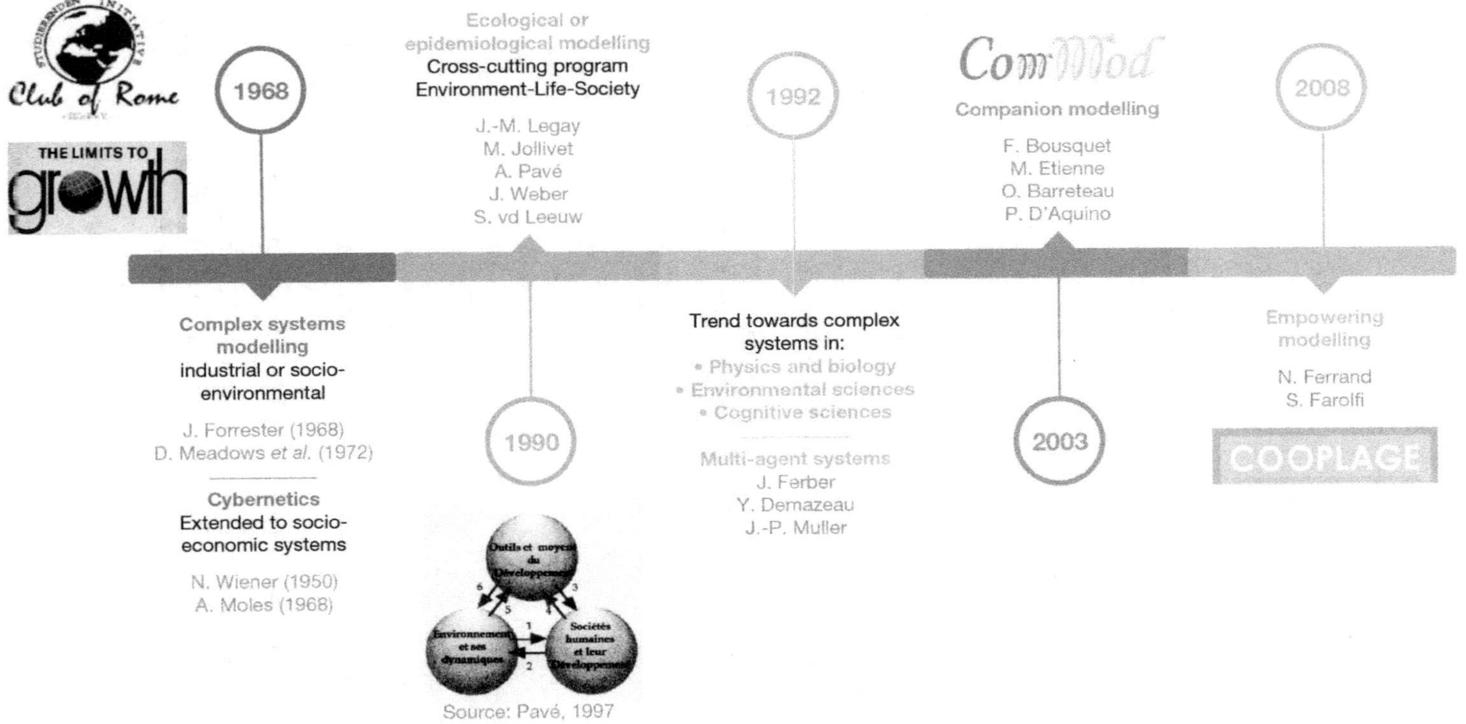

Figure 2.2. Historical background behind the CoOPLAGE approach

Box 2.1. (next)

orientation in physics and biology (dynamic systems and chaos, cellular automata, networks, percolation, renormalisation), and on the other hand, the aforementioned environmental sciences (with a growing link to geography via D. Pumain and L. Sanders), and lastly the emerging cognitive sciences between connectionism, artificial intelligence and evolutionism.

In France, since 1992, these reflections very quickly benefited from a specific contribution from research on multi-agent systems (J. Ferber, Y. Demazeau, J.-P. Muller), be it in modelling, simulation or problem solving. Multi-agent systems are a set of computer processes that run simultaneously. They allow several agents living at the same time, who share common resources and communicate with each other, to be simulated (adapted from Bousquet *et al.*, 1999). By facilitating a more natural and direct description of entities and dynamics, these individual-centred models have improved the dialogue with non-expert actors. Finally, model linking has required new thinking on the exchange of viewpoints, their dynamic implementation and adequate formalisms, and more broadly on the production and use of knowledge through modelling.

This is the basis upon which F. Bousquet, M. Etienne, O. Barreteau, P. D'Aquino and others initiated "Companion Modelling" (Etienne, 2011). Companion Modelling (or ComMod) aims at bringing different stakeholders to gradually get to know each other, exchange their arguments and viewpoints in order to build a shared view of an issue (a model) and jointly develop an accepted solution. The main ComMod methods and tools are role-playing, multi-agent modelling and social simulation. The ComMod approach is therefore an original way of approaching modelling, which is often used to support collective decision-making processes concerning the sustainable management of renewable natural resources. The approach gives non-scientific actors a role in the co-production of models. The modeller-facilitator role is central, as this person is the mediator of the various perspectives and the delivery doctor* of a common model. This requires specific expertise and strong intervention, which at first seem contradictory to the objectives of autonomy and social dissemination. From 2008, a complementary perspective put forth by N. Ferrand and S. Farolfi has provided a change in scale in Companion Modelling and has broadened its effects. This has led to the principles and tools of "empowering modelling" and to the foundations of the CoOPLAGE method of letting actors do as much as possible on their own, while facilitating their collaboration through adequate meta-models.

* In the sense of 'bringing to life'.

▸▸ Complementary postures for innovative engineering

The specificities and tools of the CoOPLAGE approach

Having acknowledged the capacity of all actors[3] to produce, formalise and compare their knowledge in structured models that can be used together, we sought to gradually empower these actors by freeing them from the facilitator. To achieve this, three concomitant constraints or objectives were taken into consideration:
– the materials (language, method, hardware kit, software) to guide them step by step in their process;

---

3. Including illiterate populations, through the use of appropriate materials.

– sufficient control through these materials to verify the quality of the model produced, in relation to the actual knowledge of the concerned sectors (water, environment, economy, etc.);
– the possibility of using the produced models for knowledge or decision-making needs, with and for the users, for example through social simulation (role-playing) or computer simulation.

In addition, whereas the body of works on Companion Modelling has focused on the dynamics of socio-environmental systems, their resilience and adaptation, CoOPLAGE sought to model other target systems or issues, based on the real needs of stakeholders. We detail these variations below. In practice, this meant proposing modelling kits, i.e. material for table-top work, accessible to all and which allow for acceptable models of the territory to be collectively established. These models can then be used to explore different transformation options and their consequences through simulation.

This led to the development of the Wat-A-Game set of tools, more specifically to the basic INI-WAG kit, and its multiple thematic and territorial variations (figure 2.3 and see chapter 12). A watershed model can be built using these tools. Various elements represent the river, its tributaries, fields, towns, forests, as well as the territory's dynamics (for instance hydrological and financial represented by circulating different coloured beads),

Figure 2.3. The Wat-A-Game Tool: A role-playing game to be built and played collectively

various actors (using role cards) and the activities they carry out there (using activity cards chosen by the players). Once constructed, the role-playing game allows the players to explore different possible transitional paths (for example by changing the activities carried out by the players or by testing the consequences of a specific event in the game). The tools in the Wat-A-Game family provide a common framework with reusable elements (a lexicon), rules (a grammar set) and a protocol to be followed together. Variable levels of modelling are proposed, from the simple reproduction of an existing model, to the mapping of a system and finally the independent production of new model elements (activities, roles, resources). From an initial model oriented towards quantitative water management, users can, for example, add quality or biodiversity issues, or add new roles. A variety of experiments have been set up using INI-WAG, including "Eau en Jeu®" (an educational kit on integrated water management for schools[4]), "L'Eau en Têt" (see chapter 13), WasteWAG (see chapter 14) and MyRiverKit (a methodological kit to raise awareness of the concept of ecosystem services[5]).

Similarly, the CoOPLAN method for participatory planning (see chapter 17), PrePar for participation engineering (see chapter 9), JustAGrid for justice dialogue and Self-Modelling for Assessing Governance (SMAG) for governance diagnostic, are also based on participatory modelling processes of different types of systems (respectively management strategies, decision-making processes, sharing and governance rules, figure 2.1). Initially, the aim is to "get the modelling done", then to gradually minimise the amount of guidance required to "let it happen". This involves, on the one hand, rapidly training local facilitators and, on the other hand, providing manuals and "self-facilitating" materials, i.e. that participants can facilitate themselves, without having to call upon a facilitator.

This set of tools and methods form the CoOPLAGE approach. These tools are currently being digitised on the CoOPILOT platform (see chapter 8). This digitisation constitutes a further step towards empowering the actors, which, however, has not yet been evaluated from an operational standpoint.

From needs-based pragmatics to research-intervention

Whether at INRAE or at Cirad (French public research institutions having hosted CoOPLAGE development), "field" culture is fundamental. Responding to the needs of stakeholders in various countries is the focus, alongside knowing how to help stakeholders formulate these needs. In parallel, our research, by virtue of its mandate to support public policies, must also respond to two other challenges: on the one hand, to generalise what we have learned from our various experiences so that this can be used elsewhere in an independent manner (in particular to minimise the need for public intervention), and on the other hand, to produce methodological innovations through experimental approaches that can lead to designing and evaluating the performance of various approaches and tools for multi-stakeholders, multi-issue and multi-level contexts.

However, these three issues (meeting the needs of stakeholders/generalising results/ producing innovations) are often conflicting. Meeting the needs of stakeholders often

---

4. http://eauenjeu.org
5. http://www.gesteau.fr/vie-des-territoires/my-river-kit-un-jeu-de-role-pour-sensibiliser-la-gestion-integree-des

implies continuity with their perceptions and current practices, which are not always compatible with the introduction of innovations that may, on the contrary, be at odds with these same perceptions and practices. Moreover, evaluating the performance of the innovations resulting from our research, in view of their potential dissemination, would require experiments with control groups to allow the various factors involved to be controlled, as is for instance done in experimental economics[6]. However, the real socio-political decision-making contexts in which we work with a limited budget (e.g. decentralising natural resource management in Tunisia, piloting participation in water policy in New Caledonia, involving citizens in institutional river management systems in France, etc.), do not allow this type of experimentation to be easily implemented. This posture often renders fragile results compared to purely descriptive research or research based on formal experiments, but at the same time it allows for truly new methodological venues to be explored.

Thus, starting with the field's needs, sometimes in an opportunistic manner, and based on the principles of CoOPLAGE, our research-intervention frameworks have a double impact: the exploration of new methods, sometimes stabilised, and various socio-technical changes for the actors in the territories. The failures encountered (non-adoption, resistance, behavioural inertia, impact limited to the project) provide new resources for the next experiment. Supported by large-scale training, we have gradually disseminated these principles and practices internationally, with the latent hope of having a lasting impact on multi-actors decision-making practices at various levels.

Truly participating in the decision on and for oneself

Participation and decisions are too often separated. Participation is too often used to facilitate the acceptance of decisions by different actors (see chapter 6). In which case, participation is restricted to communication aimed at convincing the "public" to welcome a project decided elsewhere ("acceptology"). In France, the 2016 ordinances on environmental dialogue seek to correct this by bringing the requirement for participation to an earlier project stage, so as to first discuss the opportunity, then the options and their implementation (see chapter 4). But the distribution between open, citizen participatory processes, technical and administrative appraisal, and political choices remains very unbalanced, backed by arguments concerning time, capacity and socio-economic risk (no politician wants a project with a private sector pre-agreement to be called into question by citizen participation). There are many decision-making stages for which the choice of involving these stakeholders is never made explicit or contested. Who frames and initiates a consultation for a project? Who should decide on the decision-making process? Who should participate in the diagnostic? Who can discuss "what is right"? Who can propose actions and plans? Who votes and chooses? Who implements? Everyone is involved, but there is little space to modify the roles.

As part of our experiment on support methods, we have therefore tried to ensure that the actors themselves question the place of each and everyone in the decision. This was achieved in particular through publication of the PrePar framework with support from

---

6. This would involve, for example, comparing a group that has tested an innovation with another group with similar characteristics that has not tested the innovation. Along these lines, the work in development economics that is best known to the general public is that of Esther Duflo, who received the so-called Nobel Prize in Economics in 2019.

the Rhône-Mediterranean-Corsica water agency[7]. It is based on a reference framework with eight decision-making stages (downloadable from http://frama.link/RMCPart). For each stage of the decision (diagnostic/definition of objectives/planning, etc.), stakeholders can define the desired degree of participation (low/medium/high) and then choose the appropriate participatory methods (Hassenforder *et al.*, 2020).

Testing the involvement of new actors in a decision obviously requires that they be able to do so effectively, be it in terms of capacity, resources or legitimacy. This is why, apart from a general methodological inventory, we have also sought to provide solutions to stages that have not been dealt with much elsewhere: for example, by exploring how to get people to participate in the construction of a participatory observatory (and not in the observatory itself), (see chapter 16), how to discuss and co-organise participation on a large scale, how to reintegrate monitoring and evaluation into participation to make it an asset rather than a constraint (see chapter 10), or how to mobilise digital technology to monitor the process, beyond electronic debate (see chapter 8). The aim lies in re-legitimising and putting into action the stakeholders, including citizens, in stages that are generally occupied by managers and specialists and, in this way, creating co-engagement and long-term efficiency.

Questioning, monitoring and evaluating "multi-impacts": reflexivity on change at the heart of empowerment

Firstly, the challenge of empowerment reflects the need to decentralise and minimalise intervention by public authorities. In the long term, the aim is to support the most appropriate mechanisms for developing "strong resilience"[8] locally, i.e. the capacity of stakeholders who share territories and common environmental goods to choose their future, to control their resources and to steer their dynamics, with minimised external intervention, particularly public aid and regulation. An additional methodological challenge is the fact that the various groups of actors have varying levels of conditions to resilience, which are interdependent to some extent. From this angle, the primary challenge is to help stakeholders define what they want for themselves and their environment, the acceptable pathways to achieve this, and to enlighten them on the dynamics that will allow them to evolve towards these objectives. Without prejudging their ability to choose efficient strategies (which is the subject of other CoOPLAGE tools), they must at least know where they stand and where they want to go. But any and all action has multiple environmental and social impacts, both direct and indirect.

Since the launch of the "ENCORE" (External / Normative / Cognitive / Operational / Relational / Equity – Ferrand and Daniell, 2006) monitoring-evaluation framework, we have sought to qualify all of these impacts in a global manner: whether they be transformations induced by the actors on their environment, normative changes (e.g. in values or preferences), cognitive learning, changes in practices and concrete behaviour, or

---

7. In the scope of the 'What participatory strategy for local water management with citizens' project (2016-2020).

8. Resilience in its classical definition (Botta and Bousquet, 2017) for socio-ecological systems refers to *'the capacity of an ecological and social system to absorb or withstand a disturbance or stress, while maintaining its structure and functions through processes of self-organisation, learning and adaptation'.* As the authors mention, we are more in a 'development' perspective that targets the most vulnerable as a priority (Ferrand *et al.*, 2014).

changes in relational structures and social justice. It is not just a question of observing these impacts "from the outside", with an analytical aim, but of making the actors themselves "take into account" what is changing. These "multi-impacts" are certainly difficult to measure, especially all of them, even more so "from the inside", i.e. by the actors themselves. Nevertheless, the fact that they are taken into account by the stakeholders themselves, and the fact that structured dialogue is taking place on these themes, already guarantee that what is deeply and durably at stake for them is highlighted.

Here again, modelling is at the core of our approach: the ENCORE framework and associated approach (Hassenforder *et al.*, 2016) allow actors to collectively model the desired changes and reflect together on the paths to achieving them. To this end, we are currently working on the principles of "endo-evaluative participation". The aim is to minimise the tools dedicated to evaluation (questionnaires, etc.), which are often a burden for participants to complete, and to maximise data collection on the impacts of the process through the participatory tools themselves. For example, an indicator on the strain or solidarity created between participants can be added to a role-playing game, in order to evaluate relational impacts through a simple and non-disruptive methodological adaptation. In parallel, this ambition of endo-evaluative participation is also expressed through integrated and adaptive thought on both the evaluation and the engineering of the process. In simple terms, the aim is to reflect on the changes desired, and think about how to achieve them, then to evaluate whether these changes are being achieved, and if necessary adapt the process if they are not. In any case, a major focus of our work is placed on making the participants themselves think about monitoring and evaluation as well as the engineering of the participatory approach. This is done in particular by setting up pilot groups including citizens (see insert 3 in chapter 17). This approach is quite different from classic analytical scientific approaches that advocate the independence of the evaluation stage. Reflexivity and change control are what take precedence here.

Co-adapting practices and policies: planned engineering or DIY along the way?

Most of the requests we receive are from public authorities. In general, we are called upon to help a pilot group to design and organise a participatory process that includes stakeholders at very different levels (ministers, elected officials, administrators, economic actors, experts, researchers, associations, locals, the socially excluded, etc.). The initial aim of a certain number of these requests is acceptability of a decision: in other words, for the decision-makers and pilots of the participatory process, the objective is to get a decision accepted, for example the creation of a new reservoir or the implementation of a new regulation. As researchers and for those who facilitate the process, our goal is then to help make this request evolve towards a vision of co-construction and co-evolution, i.e. to make the pilot group and decision-makers understand that involving other actors in a decision that has already been taken is of little or no interest. To accompany this evolution, it is important to get the groups to ask themselves a certain number of questions related to the organisation of participation: which roles should be given to which actors (pilot, reference person, participation warrant, facilitator, observer, etc.)? What should be imposed and what should be discussed in the participatory process? What materials should be used? What training is needed? Should an external facilitator be hired or should someone be trained internally? How should sub-groups of participants be organised? How can "silos be broken down"? How can trust be rebuilt? etc. In trying to answer these questions, the pilot members of a participatory approach often find themselves in a "do-it-yourself" (DIY) posture based on empirical know-how and observation rather than on systematic theories. This only works within the limits of the pilots' skills, hence the need to call upon experts.

To overcome this "DIY" stage, different options need to be tested in different contexts and with different actors for each of these questions in order to analyse which options are the most relevant to the final objectives. This is what we seek to do by conducting comparative analyses of the different participatory approaches we support. This has also led us to propose a "meta-model" for participation engineering, i.e. a model that can be used for different participatory processes, in different contexts and with different actors, and which can be transferred to any pilot group to enable it to design and implement its own participatory process quickly and with minimal support.

This meta-model is the PrePar method, mentioned above and presented in the chapter 9. PrePar proposes a participation engineering framework centred on systematic deliberation of the forms of involvement of all actors at each stage. Participants are asked to define the actual roles of the different actors in the successive decision-making stages. The method thus allows for a participation plan to be produced, in principle, and details the different actions to be carried out, the participatory methods to be used and the actors to be involved. A digital version of the method (ePrePar) is available.

The deliberation carried out through the PrePar process provides the basis for drawing up a participation charter. Here again, using PrePar in a participatory way is a new approach, the implementation of which constitutes a major change in posture and supports impactful social learning: the stakeholders, including citizens, discuss the way in which they will be associated to the target decision, as well as the commitments and responsibilities of each. The subsequent adherence to the rules and results depends on this, and consequently the mutual trust between participants, regulators and organisers of the participatory process. Admittedly, however, this participatory planning "of the participatory process" has as much value as a preparatory process as the plan produced, which can be quickly adapted, modified, adjusted... Consequently, there is a real compromise between this planned engineering, structured by the meta-models in PrePar, and all the adaptive steering required later by the contingencies of the socio-political path.

►► The decision model of the CoOPLAGE support platform

The goal of the CoOPLAGE tools is therefore to accompany and coordinate the decision-making stages of actors at all levels, from citizens to elected representatives and managers, in order to facilitate technical, social and institutional changes that are compatible with environmental constraints and achieve the sustainable effects sought by the participants.

The decision model presented in table 2.1 and figure 2.4 can be used as a synthesis of all the CoOPLAGE tools and the decision-making stages at which they can be mobilised. Each step corresponds to a stage in decision-making:

Participation =
"Sharing decision process + Piloting + Preparation + Diagnostic + Prospective
(foresight) + Preferences + Planning + Prioritisation + Implementation".

These steps were initially based on the four phases of the decision-making process identified by Simon (1977) "Intelligence/Design/Choice/Review" and have been adapted to best fit the needs of stakeholders and the standard steps in elaborating water policy (for details, see annex 4 in Hassenforder *et al.*, 2021). Although CoOPLAGE allows stakeholders to reflect on all the steps in the preliminary engineering phases, only some of them are actually formalised (table 2.1).

The different approaches and tools as well as their operational implementation are presented throughout this book and in figure 2.4.

Table 2.1. The decision model of CoOPLAGE and its instrumentation (extended version)

Step	Description	Corresponding CoOPLAGE tools
Sharing	Combine face-to-face and digital means to structure and share the process between actors at all levels	CoOPILOT (digital platform containing all CoOPLAGE tools)
Piloting	Co-construct criteria to evaluate the process and its socio-environmental impacts, then monitor and use these criteria to pilot and adapt along the way	ENCORE (External, Normative, Cognitive, Operational, Relational, Equity – corresponds to the different types of impacts that can be evaluated)
Preparation	Train the actors, then co-design and organise the participation by discussing roles, commitments and methods, to obtain a consensual participation plan and charter	PrePar (to prepare and reflect on a participatory approach) MOOC Terr'Eau & co (online course for training in the CoOPLAGE approach) INI-WAG (Wat-A-Game basic kit to understand the principles of an integrated water management role-play)
Diagnostic	Observe, diagnose, understand and model the social and environmental situation	ROCK (River Observation and Conservation Kit – observation sheet to be created to observe a river or a territory) SMAG (Self-Modelling for Assessing Governance – to produce a diagnostic of the past governance of a territory)
Prospective	Imagine future scenarios, explore possible paths, simulate	CreaWAG (version of Wat-A-Game to create role-plays on integrated water management) and the so produced specialized models and games
Preferences	Discuss the goals and constraints of the actors in order to define the management framework, with a specific focus on social justice	JustAGrid (to dialogue on justice issues)
Planning	Formulate options for action, then characterise and assemble them into multi-level, feasible and efficient territorial strategies	CoOPLAN (to develop an integrated water management plan in a participatory manner)
Prioritisation	Compare and prioritise strategies in order to choose one	
Implementation	Assist in institutional (governance) and operational (technical, economic) implementation	

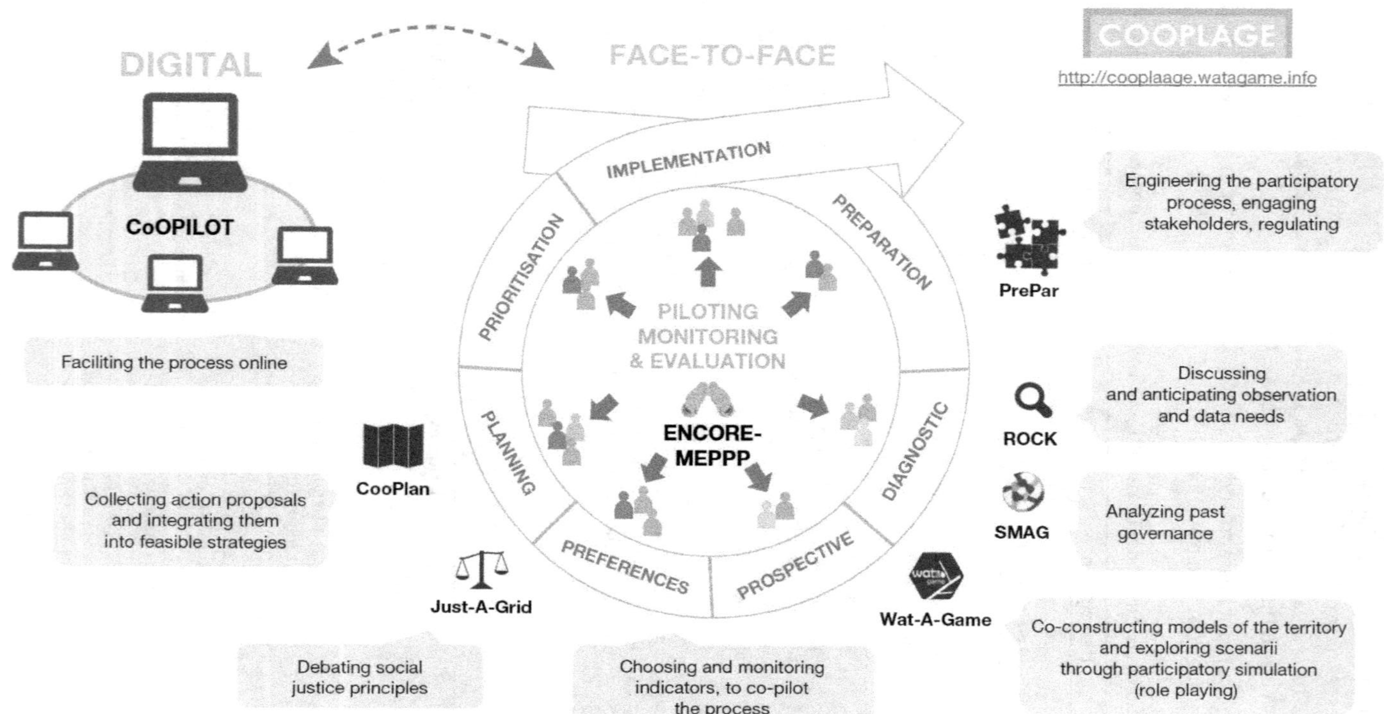

Figure 2.4. CoOPLAGE decision model and its instrumentation

▸▸ References

Botta A., Bousquet F., 2017. The resilience of social and ecological systems: taking account of uncertainty for development. Cirad, Montpellier, *Perspective* 43. https://doi.org/10.19182/agritrop/00003

Bousquet F., Barreteau O., Le Page C., Mullon C., Weber J., 1999. An environmental modelling approach: the use of multi-agent simulations. *In:* F. Blasco (Ed.), *Advances in Environmental and Ecological Modelling*. Elsevier.

Etienne M., 2011. *Companion modelling A participatory approach to support sustainable development* (M. Etienne, Ed.). Versailles, éditions Quae.

Ferrand N., Daniell K.A., 2006. Comment évaluer la contribution de la modélisation participative au développement durable ? *Séminaire DDT*.

Ferrand N., Hassenforder E., Abrami G., Daniell K.A., 2014. *JUST-A-GRID, when people co-model fair resource allocation*. Resilience 2014 – Resilience and Development: Mobilizing for Transformation.

Forrester J. W., 1968. *Principles of Systems*. MIT Press.

Hassenforder E., Barreteau O., Barataud F., Souchere V., Ferrand N., Garin P., 2020. Enjeux et pluralité de la participation dans la gestion intégrée des ressources en eau. *In* : M. Voltz, D. Burger Leenhardt, O. Barreteau (Eds.), *Eau et agriculture : gestion intégrée et gouvernance territoriale*. Versailles, éditions Quae.

Hassenforder E., Girard S., Ferrand N., Petitjean C., Fermond C., 2021. La co-ingénierie de la participation : une expérience citoyenne sur la rivière Drôme. *Natures Sciences Sociétés*, 2, 159-173. https://doi.org/10.1051/nss/2021050

Hassenforder E., Pittock J., Barreteau O., Daniell K. A., Ferrand N., 2016. The MEPPP framework: A framework for monitoring and evaluating participatory planning processes. *Environmental Management Journal*, 57(1), 79-96. https://doi.org/10.1007/s00267-015-0599-5

Meadows D. H., Meadows D. L., Randers J., Behrens W. W., 1972. *The Limits to Growth: A Report for the Club of Rome's Project on the Predicament of Mankind*. Universe Books.

Minsky M., 1965. Matter, Mind and Models. *Proc. of the IFIP Congress*, 45-49.

Moles A., 1968. Cybernétique, information et structures économiques. *Les Cahiers de La Publicité*, 19(1). https://doi.org/10.3406/colan.1968.5034

Pave A., Jollivet M., 1993. L'Environnement un champ de recherche en formation. *Natures Sciences Sociétés*, 1, 6. https://doi.org/10.1051/nss/19930101006

Schmidt-Lainé C., Pavé A., 2002. Environnement : modélisation et modèles pour comprendre, agir ou decider dans un contexte interdisciplinaire. *Natures Sciences Sociétés*, 10(1), 5-25. https://www.nss-journal.org/articles/nss/pdf/2002/02/nss200210sp5.pdf

Simon H. A., 1977. *The New Science of Management Decision, Revised Edition*. Prentice-Hall.

Wiener N., 1950. Cybernetics. *Bulletin of the American Academy of Arts and Sciences*, 3(7), 2-4. https://doi.org/10.2307/3822945

Wiener N., 2019. Cybernetics or Control and Communication in the Animal and the Machine. *In*: *Cybernetics or Control and Communication in the Animal and the Machine*. https://doi.org/10.7551/mitpress/11810.001.0001

Chapter 3

The context of citizen participation in water management in France

Audrey Massot, Anne Pressurot and Marie Trouillet;
interviews conducted by Emeline Hassenforder

> Ever more land and water managers are implementing participatory approaches in the scope of their projects, plans and programmes. This chapter provides elements for understanding the context surrounding citizen participatory processes in France. It is based on the testimonies of three actors, each of whom sheds light on the subject from her own perspective, altogether encompassing the local level, the regional Rhône-Mediterranean-Corsica watershed level as well as the national level.

The aim of this chapter is to provide elements for understanding the context surrounding citizen participatory processes in France. It discusses:
– current trends, in particular through the French ordinance for the democratisation of environmental dialogue;
– key events such as the conflict surrounding the Sivens dam project or the introduction of the Gemapi tax[1];
– main obstacles and levers, such as the willingness of elected officials or the articulation between different territorial policies;
– and finally, a few anecdotes and recommendations to those who are in charge of future participatory approaches.

It should be noted that this chapter deals mainly with participatory approaches that support the development or revision of public policies. More "spontaneous" participatory processes (demonstrations, petitions), those carried out exclusively by civil society actors or those that put action first (living labs or citizen initiatives and forums) are not discussed.

---

1. The Gemapi tax for the management of aquatic environments and flood prevention (Gemapi or Gestion des milieux aquatiques et de prévention des inondations) is an optional tax that can be imposed on private individuals or legal entities since 1 January 2018 by municipalities or public establishments for intermunicipal cooperation, which are the competent authority in terms of aquatic environments and flood prevention. The tax aims at financing actions related to this new competence: development of watersheds, maintenance and development of watercourses, canals, lakes and other water bodies, flood prevention mechanisms, protection and restoration of wetlands, as well as hydraulic installations and their maintenance (Source: https://www.senat.fr/questions/base/2018/qSEQ180906795.html, https://fr.wikipedia.org/wiki/Taxe_pour_la_gestion_des_milieux_aquatiques_et_la_pr%C3%A9vention_des_inondations

The floor was given to three actors, each of whom sheds some light at her own level:
– At the local level, Marie Trouillet (M.T.) is a facilitator within an association called the Centre for Environmental Initiatives CPIE[2] in Bugey-Genevois (east of France, at the border with Switzerland). She has been supporting participatory approaches in favour of the environment in the Haute-Savoie area for the past eight years. The testimony gathered represents Marie Trouillet's perspective and does not speak for the CPIE Bugey-Genevois in any way.
– At the Rhône-Mediterranean-Corsica watershed level, Anne Pressurot (A.P.) of the Rhône-Mediterranean-Corsica water agency, formerly in charge of evaluating the agency's public policies and research projects on participation and elected officials, is currently intervention officer at the Regional Delegation in Lyon. The comments made here are the sole responsibility of Anne Pressurot and do not commit the Rhône-Mediterranean-Corsica water agency.
– At the national level, Audrey Massot (A.M.) is in charge of the territorial coordination of the water policy at the Directorate for Water and Biodiversity of the Ministry of Ecological Transition (MTE[3]). The statements made here represent the views of Audrey Massot and do not commit the Ministry of Ecological Transition.

▸▸ What are the current trends in participation in the water sector?

A.M. (Ministry): At the national level, there are currently three major factors or trends that influence participation. The first of course is the 2016 Ordinance on the Democratisation of the Environmental Dialogue. It requires water managers, in particular for water management plans (SAGE[4]), to either engage upstream in a consultation or to produce a declaration of intent including a right of initiative (see chapter 4). A public consultation has to therefore be included as early as possible in well-established procedures such as the SAGE procedure, which for instance has been in existence for 25 years now. Some regions appreciated the advent of this ordinance and had already consulted the public during the preparatory phase of their SAGE (e.g. the Assises du Loiret launched in 2016[5], which consisted of a photographic survey that was carried out as part of the Calavon-Coulon SAGE). In other regions, however, it was seen as an additional regulatory phase and therefore something that made the procedures even more tedious. At the ministry, we are working on the methodological and procedural framework of this consultation alongside other stakeholders and the National Commission for Public Debate (CNDP[6]).

This democratisation of environmental dialogue echoes a second trend which consists in the mobilisation of citizens at all levels and in many forms. Some mobilisations are

---

2. Centre permanent d'initiative pour l'environnement (CPIE)

3. Ministère de la transition écologique

4. SAGE – Schéma d'aménagement et de gestion de l'eau: It is a planning tool, instituted by the 1992 French Water Act, aiming at the balanced and sustainable management of water resources at the watershed or aquifer level. The SAGE sets, coordinates and prioritises general objectives for the use, development and quantitative and qualitative protection of water resources and aquatic ecosystems, as well as the preservation of wetlands. It identifies the conditions and means for achieving these objectives (Source: https://www.gesteau.fr/presentation/sage)

5. For more information in French: http://www.assises-riviere-loiret.fr/index.php

6. Commission nationale du débat public

highly visible and covered by the media (climate marches, zones to defend); others are more discreet but just as important (think tanks, local associations such as WARN![7], the "time to question" citizens' questionnaire broadcast by Arte as well as several associative movements in 2020, etc.).

Lastly, a theme that I personally have noticed on the rise is the issue of water quality, which usually mobilises the general public more than quantity. Topics such as glyphosate or plastic pollution generate greater awareness, perhaps because the risks are better understood. These are not new issues, they were already being talked about 30 or 40 years ago, but they are back on the agenda and worrying the population. So, it's good that the general public is paying attention. The quantitative aspect, on the other hand, is still often reserved for scientific and technical stakeholders, and for economic actors directly impacted by water deficits. There are initiatives on the water management side such as the study of collectable volumes on the Roussillon aquifer or the regional water management project on the Usses but I have the feeling that it's more difficult to mobilise the general public on the issues of drought and low water levels. It may be because many people believe that drought mainly affects the south of France, whereas other basins, Seine Normandy or Artois Picardy, are equally affected.

A.P. (Water agency): One of the important trends in my opinion is that increasingly more citizens and locals are being included in participatory approaches for water management. Consultation has always been an operating principle with multi-party working committees or commissions allowing all stakeholders to express themselves (see the 1992 Water Act), but it has mainly concerned representatives of associations, administrations, companies or elected officials. The direct participation of citizens and local residents in participatory processes is more recent. This evolution has been notable throughout the implementation of the 10th action programme of the water agency (2013-2018), which is more oriented towards land management planning and therefore producing greater impact, as well as through the 2016 national Ordinance on Environmental Dialogue. Incidentally, the water agency also signed the Ministry of the Environment's charter on public participation at the end of 2016 (see chapter 4).

This trend is reinforced by the fact that many elected officials are more open to participation than they were before. They have understood the importance of taking the environment into account in public policies and are trying to open up the reflection on water projects to citizens by organising consultations and debates. At the same time, more and more citizens want to get directly involved without going through their representatives; they are more active and vindictive. This is the case, for example, of the counter-urbanisation movement in which people settle in the countryside to be closer to nature and therefore defend the landscape, biodiversity and the fact that they can walk along the water. These new subjects, such as the emotional and sensitive relationship with rivers or the restoration of watercourses, are often complex issues and therefore require participatory approaches to obtain a consensus at the local level.

---

7. The 'WARN !' movement (We are ready now !), which regroups activists and to a certain extent whistle-blowers, was initiated by a group of youth who had participated in the Conference of Youth (COY11) at the Conference of the Parties on Climate in 2015 (COP21). Today, this movement sets up workshops on the environment in schools and organises large-scale awareness programmes to sensitise the public-at-large on global warming and the ecological emergency. http://wearereadynow.net/

M.T. (Association): For me, there has been a change in the level of participation. In the past, we mainly built up awareness on the ground, for example by encouraging people to reduce their water consumption. Whereas now, participation has moved into other spheres; we are asking participants to give their opinion on water management, on the quantity of water or on well-functioning mechanisms. We are called upon for other things than just awareness. People are no longer just informed, they can also express their views on a wider range of issues than before.

At the same time, there is less and less support for project managers or structures carrying out participatory processes or for elected officials in implementing participation. Yet, I've seen that the people who benefited from the support of programmes like *Osons Agir*[8] had developed a real sense of participation; this was a true lever for the success of the participatory approach. However, today, those people who obtained support in certain territories are no longer there, following elections or a change in jobs, and the support programmes have for the most part been replaced by the occasional training days. So the trend could well be reversed.

▸▸ For you, what have been the main developments in participation in the water sector in recent years?

A.P. (Water agency): The conflict around the Sivens dam project[9] was an electro-shock in the water sector on the importance of involving citizens. It led to major legal changes, in particular to the integration of a prior consultation into SAGE policies on water management. It also affected all land use planning and river restoration projects.

The second influence for me is the reduction in the amount of available water due to climate change, as well as the pressures on uses which are increasingly strong. These pressures reinforce the need to exchange, participate and agree on the sharing of water resources. In some territories of the Rhône-Mediterranean-Corsica basin, these pressures have created tensions between stakeholders, which have even required mediation measures.

The impoverishment of society, further marked by the Covid-19 crisis, has also pushed certain issues up the agenda. This can be seen, for example, in the social pricing of water or the safeguard of low-income families from water cut-offs. We realised that there are sectors where public services have to be ensured and where the economy-water-common good link is essential.

Lastly, the flooding of the Grand and Petit Travers coastal dunes in the Hérault department (hemmed in between the Étang de l'Or and the sea and located between Carnon and La Grande-Motte in South of France) marked participation as this created conflict between citizens who were for or against certain urban developments, in a context where all involved were extremely sensitive and impulsive on the issue. In many cases, participation was a means of calming conflictual situations.

---

8. 'Osons Agir' is a programme carried by the Regional Union of CPIEs in the Rhône Alps region - Union Régionale des CPIE Auvergne Rhône-Alpes. It aims at helping professionals, elected officials, and citizens to build their skills in participatory approaches. It offers group workshops as well as personal learning sessions. http://urcpie-aura.org/nos-missions/accompagner-les-territoires/dialogue-territorial-osons-agir/

9. In 2014, a dam project on the Tescou river in the Garonne basin led to violent clashes between activists opposed to the project and anti-riot police. These clashed led to the death of an opponent in October 2014. The dam project was abandoned in December 2015 by prefectural decree.

A.M. (Ministry): There are two examples of territories where citizen mobilisation has been particularly publicised. The Sivens dam of course—which by-the-way was the event that triggered reform on environmental dialogue in 2016—but also the gold mountain in French Guiana. This mining project has left its mark due to the very strong mobilisation of local communities. It was a complex project, like all mining projects, with strong environmental issues related to water, wetlands and forests, along with strong economic implications as well. The project was highly publicised and politicised. In the end, it was abandoned.

Aside from these already well-known events, many requirements have already been written into law mandating the public to get involved in a timely manner. At the Ministry level, consultations in the framework for water installations, planning and management (SDAGE[10]) also constitute significant events. Every six years, the public is consulted on the important issues to be addressed in the SDAGE as well as on the work programme. Some of the participatory sessions have been very useful and successful, for instance in the Martinique basin.

Lastly, at the European level, the "Fitness Check" questionnaire was sent to all European countries in view of a possible revision of the Water Framework Directive. Citizens can answer the survey directly. And even if the questionnaire is in English and relatively technical, it allows concerned citizens or organisations to give their opinion and to be heard by the European Commission.

M.T. (Association): The introduction of the Gemapi tax had a certain impact at the local level because people realised that their bills had increased, whereas their consumption had not. However, this only affects certain people; most people are disconnected from these changes, and do not even realise them.

On the other hand, what has significantly marked participation is the trend towards the grouping together of districts in the form of greater municipalities or public establishments for water planning and management (EPAGE – Établissements publics d'aménagement et de gestion des eaux). These groupings have created a disconnection between water managers and the population. People used to know those who were in charge of water in their districts, and the managers came to see them directly if there was a leak or other problem. There was a dialogue and people felt concerned by the water issue. But now, even the elected officials are disconnected from this because the competence has gone to the greater municipality or elsewhere. So, even if these groupings are of interest for territorial strategies or watershed solidarity, they have created a disconnect between water managers and the population.

▸▸ In your opinion, what are the main obstacles to implementing a participatory approach today?

M.T. (Association): One of the primary obstacles is the lack of time that project officers dedicate to participation. Their time is essentially devoted to large technical investment projects and participation is ranked second, even third or fourth in terms of priorities. Participatory processes can be time-consuming, especially if they have never been set up before. Further, officers do not always think of forming a partnership with a local structure or signing an agreement with an association or other organisation.

---

10. Schémas Directeur d'Aménagement et de Gestion des Eaux

Another obstacle lies in the articulation between the different territorial policies. Today, the primary tools used for land management—such as regional frameworks on cohesive land management (SCoT[11]) and local inter-urban planning (PLUi[12])—do not leave room for citizen participation and dedicate even less space to water. These tools take participation into account through association or union representation; citizens are only consulted, and are rarely directly involved through constructive group workshops. In one of the catchment areas I work in, the local citizens had told the river committee that, in their opinion, one of the priorities for dealing with the lack of water was to reflect on land management policies. In this basin, new settlements have increased the pressure on resource sharing. Elected officials told them that this issue would be addressed in other instances (in this case the PLUi) in which citizens do not have the opportunity to express themselves collectively. As a result, citizen participation in water management often only leads to awareness campaigns or to small investments such as water collection systems, but not to profound changes in land management.

A.P. (Water agency): The main obstacle is primarily political, and lies in the willingness of elected officials to set up participatory approaches or not. Everyone's role must be clear: who decides, who discusses, until when, on what, without restricting anyone's expression, whether they are well or just a little informed on the subject. The second obstacle lies in the competency, know-how and interpersonal skills needed to implement participatory approaches. Participation requires expertise, particularly at the social level, to lead and mobilise a large and representative audience. Not everyone can do this. The third obstacle, by far not the least, lies in the difficulty of rendering the process and its results transparent, and of giving feedback to the participants on what their participation has produced and what influence the participants have had on the decision, project, plan or programme. If this feedback is not provided, participants may be led to believe that they have participated in an "alibi" process. Transparency is a strong lever.

A.M. (Ministry): The relatively heavy administrative and regulatory burden imposed by legislation on managers in terms of public participation. We are aware of this. And although involving the public is meant as a good intention, we understand that it imposes fairly lengthy procedures, which can demotivate certain structures. Especially since managers still lack support on methods and training for participation. The French Biodiversity Office report edited by Contrechamps in 2018[13] has clearly identified this. It is with this in mind that we are working here at the Ministry along with INRAE national researchers, the CNDP and guarantors, investigative commissioners, and other relays in order to increase the power of these methodological levers.

Another obstacle is that it is perhaps more difficult to imagine new participatory methods for plans and programmes that have existed for a very long time and which have well-established procedures that technicians and managers are accustomed to implementing. By imagining slightly more flexible practices, things could probably be rethought and invented to ensure participation throughout policy-making, to integrate

---

11. Schémas de Cohérence Territoriale
12. Plans Locaux d'Urbanisme Intercommunaux
13. Chémery, J-B., Gasc, G., Arama, Y., Dubois, N., De la Rocque, J., Renoullin, M., Assessment of participatory approaches to integrated and sustainable water and aquatic environment management – Final Report, July 2018 – in French: http://www.gesteau.fr/sites/default/files/rapport - etat_des_lieux_gestion_de_leau_et_des_milieux.pdf

citizens from the beginning of the process up until its implementation. Local water commissions, for example, are admirable consultative bodies, and have been in place for quite a while now. Some local water commissions are already thinking about involving citizens, perhaps without giving them a decision-making role (no voting privileges which the commission members have) but simply allowing them to be part of the discussion (e.g. SAGE Drôme, Scarpe aval or Scarpe amont, SAGE Clain, SAGE Charente, etc.). The two approaches are compatible.

▸▸ How do you see the future of participation in the water sector?

M.T. (Association): Very optimistic! I really get the idea that water like the environment are becoming more cross-cutting issues that will be reflected in all areas: regional planning, the economy, health, etc. And that, thanks to citizen participation.

But for this transformation to be effective, the quality, rather than the quantity, of implemented participatory processes must be improved. There are still too many citizen participatory processes in the water sector that end in a "crappy water" (if I may say so); a three-year participatory process is carried out and then, in the end, nothing that was proposed by the citizens is implemented. As a result, people are less and less inclined to participate. These processes are already time-consuming for them, and the time taken by citizens is not the same as that of the managers or public authorities. We therefore need better participation, which is commensurate with the energy we put into it, which shakes things up, and which uses available means in a more intelligent way. We need to provide the necessary resources for the post-participation process, so that actions can be implemented.

The problem is that the concerns of citizens do not necessarily correspond to the concerns of managers or the water agency. The population are moving the lines and these lines are not necessarily in phase with the initial budget lines. And rather than seeing this as a hindrance, I think we should see it as an opportunity; citizens can provide a link between the various territorial policies because they do not feel limited by a particular field of competence or policy area as managers or elected representatives might. And for me, citizen participation will be a driving factor that will allow for water to be taken into account more in regional planning, development and life. Many water authorities exist, and yet water is still not fully considered in regional planning: we continue to build on marshland, and when housing is built, no one asks how we are going to supply it with water, or how collection systems will be set up. When we approach the environment in a compartmentalised way, we don't deal with the real issues. Citizen participation can help us get back on track.

A.M. (Ministry): In a positive way because environmental, ecological and climate change issues are beginning to make their way into people's minds and into politics. I think that the climate change approach will succeed in mobilising the general public because the climate, which is highly publicised in the media, speaks to the greatest number. With more support in methodology for participatory processes in climate change, water managers will enhance their skills. And since the climate is a highly politicised subject, elected officials will become increasingly involved and thus involve their electorate in climate issues. I think that at some point the electorate and the public in general will get involved in water issues. The political sphere and the public sphere will come together.

I'm also confident in education and the role of schools and higher education. More and more curricula include courses on the ecology, the environment and water resources. I studied at the ENTPE[14] institute for public works, an engineering school that was originally focused on civil engineering. For the past ten years or so, the curriculum has focused much more on environmental issues, with courses specialised in the management of waterways and the coastline. It is also through education that we will make citizens aware that they need to play a role in water management. I therefore believe that participation in consultations will increase.

Lastly, one of the future challenges, it seems to me, is carried by the water agencies, since they are the most visible to citizens due to the fact that they send out the water bill. It is important that citizens know what this money is going to be used for. This materialisation through cost is important. It's a sort of general public contribution: I am paying something to protect our water resources. It makes people realise that water is not a free or inexhaustible resource, and that they must therefore contribute to its preservation, be it by participating in consultations or by paying their bill.

A.P. (Water agency): On the large basin scale, the Rhône-Mediterranean-Corsica water agency is promoting societal debates to help imagine the upcoming challenges for future frameworks on water management (focus groups) or the agency's future policy, for example, on drug residues or nanoparticles in water, the use of treated wastewater for irrigation, etc. In addition, several research and development projects have been set up or are underway to ensure that participation and consultation are well articulated in the water sector[15]. The notion of common good instituted by the 1964 water law is a foundation and a plus for the development of participation in water policies.

At the local level, I imagine a very operational and pragmatic use of citizen participation to provide more substance and hindsight to local water commissions or river committees on blocked or new issues (Gemapi, sharing of water resources in territories under stress, new SAGEs to be written, reviving citizen appropriation of territories, etc.). Different commissions are often created with the same representatives when it comes to organising the water sector. Participation will breathe new life into this organisation.

Lastly, it seems to me that participation should be transformed into a more ad hoc and timely approach with easy-to-use, readily available tools. The focus is still on institutional participation and expectations are high, whereas the water sector already has consultative bodies and a very strong logic of inter-actor participation. What is lacking is participation that is more open towards citizens and set-up according to specific needs (monitoring of water resources and biodiversity, fight against heat islands in the city, restoration of waterways). The training-action plans offered by organisations or consultancy firms (such as the Centres permanents d'initiatives pour l'environnement, France Nature Environnement, etc.) help strengthen the capacities

---

14. École nationale des travaux publics de l'État

15. This is the case, for example, of the project 'What participatory strategy for local water management with the citizens' (2016-2020) financed by the Rhône-Mediterranean-Corsica water agency and led by the National Institute for Research in Science and Technology for the Environment and Agriculture (Irstea), which became INRAE in January 2020 (UMR G-EAU) https://frama.link/RMCPart

of those who support the implementation of these types of approaches. Moreover, depending on the issue, the skills in the field, the ability to act, there may be a multitude of participatory methods that are more or less costly, creative, integrated into the decision-making process...

⇥ Do you have any particular anecdote, anything that stands out, that you would like to share?

A.P. (Water agency): Yes, it was during a participatory workshop in the Drôme department. I was struck by an elected official who came to realise what an important contribution participation plays in processes. He said: "I was worried. I didn't know what I was getting into and in fact it brought me closer to the people; they took their territory into their own hands and in the end, it made relations more fluid". Especially since the approach generated a lot of citizen proposals and required a lot of investment from the managers. So, the fact that afterwards this elected official said that it had brought him a lot of new ideas, that he was no longer afraid of participation, I found that very strong.

M.T. (Association): During a forum theatre workshop[16] with locals from the Usses watershed, there was a scene where a child was wasting water because he was having fun with it. And overall the audience was very uncomfortable; they didn't know how to react because they didn't want to stop the child from having fun with the water, and at the same time they were aware of the waste that it generated. We were at an impasse. At this point, a person who had taken part in previous workshops came on stage, bringing a fresh perspective as always. She didn't explain what she wanted to do, she simply took the child by the hand and suggested that he play with something else. That made an impression on me because, for me, the answer was there: it's not a question of forbidding the different actors to do this or that, it's a question of finding the right answer for each individual and of doing things differently. It's not a question of telling farmers: you have to water your corn less; you have to see if you can cultivate differently, with another variety that is more resistant to drought, another irrigation technique or another crop. It's not a question of saying don't do it, we must collectively do it differently.

A.M. (Ministry): Yes, the public consultation in Martinique to revise the SDAGE framework on water installations, planning and management 2022-2027. The consultation was carried out in a fun way, with a travelling device that mobilised six pairs of young Martinicans, named the Blue Ambassadors, who travelled around Martinique to meet the general public to collect the population's opinions. They conducted thousands of questionnaires on drinking water, rivers, mangroves, and the results of previous SDAGEs. Several questions dealt with the trust and satisfaction of the population in the drinking water, which is a major issue in Martinique. There was a high level of participation. This water basin obtained the most responses at the national level[17].

---

16. Forum theatre is a participatory tool in which actors perform a scene illustrating a sticking point or problem between different actors. At the end of the scene, the audience is given the opportunity to replace one of the actors in order to find a solution to the problem.

17. 1.53% of the population of Martinique participated in the consultation (Source: Synthesis of the 2018-2019 Consultation of the public and stakeholders for the revision of the SDAGE). For more on the results of the consultation - in French: https://www.observatoire-eau-martinique.fr/politique-de-l-eau/cadre-reglementaire/consultation-du-public

The consultation also included online surveys. The SDAGE took the collected opinions into account in orienting its provisions. This is an example of participation that was not experienced as a regulatory constraint but rather as something highly voluntary, a way of re-involving and re-mobilising citizens on water issues and turning them into actors, in particular by mobilising young people. Young people often have a more forceful discourse on these issues today, as illustrated by Greta Thunberg[18].

►► What would you say to people who want to start a participatory process?

A.M. (Ministry): Public participation should not be seen as a regulatory phase that has to be implemented, but rather as an opportunity to enrich the plan or programme with various opinions. As a manager, we know less about the region than the residents who have seen it evolve over decades.

I would also advise anticipating the participatory process so that it goes as smoothly as possible, by assessing the forces present and any potential areas of conflict. Prior consultation is a way of defusing tensions on the territory by showing the willingness of State services and water managers to co-construct a strategy with all concerned stakeholders. Anticipating also means going to other territories to see what has been done there, identifying failed and best participatory practices. This feedback is essential, including national online consultations.

And finally, I think that the general public should be involved as much as possible, not just as a simple contributor or observer. We need to deploy methods so that citizens feel that they are actors in the process and as involved as possible. This is what we are encouraging at the ministerial level in the General Commission for Sustainable Development.

M.T. (Association): I would say that you have to "think carefully beforehand", "take time for the process", "co-construct with the participants", "see if it is legitimate", etc. But in fact, if I had to say only one thing, it would be test, experiment, don't hesitate to look for participatory tools that make people want to take part and that accompany as many as possible. When I say "as many as possible", I don't mean in quantity but in diversity. We must not only reach out to the learned, but also to people in precarious situations. Greater education and nature activities, for example, are full of tools for reaching diverse audiences.

A.P. (Water agency): That you have to be open to exchange and new things, to have confidence and trust. The world of participation is very rich; there are lots of different tools, a lot of know-how and life skills that are very inspiring. And that you should not be afraid of participation, because participatory approaches often go hand-in-hand with project endorsement or approval of planned decisions, and when this is the case, they provide for a broad and incisive perspective that legitimises the project.

---

18. Greta Thunberg is a Swedish activist who started protesting in 2018 at the age of 15 against inaction on climate change. She gained international recognition for her activism and speeches.

Part 1

The Foundations of Public Participation for Socio-Ecological Sustainability

Chapter 4

Developing a culture of participation: Progress and considerations in France

Joana Janiw; Interview conducted by Emeline Hassenforder

Whether at the international level (Aarhus Convention) or with regard to diverse French national regulations (on the environment, town planning, local authorities, etc.), direct involvement of citizens in the democratic process is presently a well-established procedure. Indeed, a review of the last few years shows public participation as an exponentially growing dynamic. This interview with Joana Janiw, Head of the Culture of Public Participation Unit at the General Commission for Sustainable Development at the French Ministry of Ecological Transition and Solidarity, takes stock of recent progress on the subject in the environmental sector in France.

▸▸ Can you explain what the recent changes have been in terms of public participation in the environmental sector in France?

The latest advances in the democratisation of environmental dialogue were introduced by the Order of 3 August 2016 reforming the procedures for informing and involving the public in the preparation of certain decisions likely to have an impact on the environment.

"Upstream" participation (public debate, prior consultation), which takes place at the development stage where all options are open, has been strengthened. Access to the right of referral to the National Commission for Public Debate (CNDP[1]) has thus been broadened, notably with the right of initiative. In addition, the device for prior consultation has been consolidated both by the introduction of minimum procedures and by the institutionalisation of a warrant appointed by the CNDP.

Finally, a new device was introduced with the conciliation procedure, and a "participation continuum" mechanism was set up to ensure a "participation log".

"Downstream" participation, which takes place after project submission at the final approval stage, has been revised. The 2016 ordinance modernised procedures by providing, for example, the possibility for a single public consultation and developing digital access to participation. The public consultation report is now systematically posted online; digital posts are open to the public. In some cases, a digital procedure for public participation, which does not involve a regulatory instance, may be organised.

---

1. Commission nationale du débat public

Four years later, it is important to review how they have fared, not only in terms of compliance with the law, but also in terms of the manner in which methods and tools were implemented to ensure that principles have been respected. This presupposes the development of a true culture of public participation and the ability to rely on what can now be described as participation engineering.

▸▸ What purpose do participation charters serve?

Participation charters can serve two main functions. They can:
– provide a basis for values and principles to which the various actors commit, so that participation can be effective and constructive;
– serve as a reference to guide the implementation of the participation process.

The Public participation charter was created in 2016 with this in mind (figure 4.1).

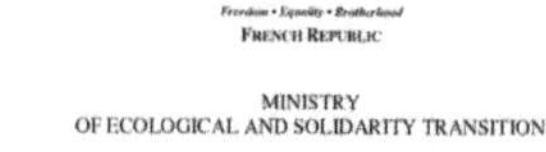

Preamble

The Charter of Public Participation proclaims that everyone must be able to participate in the elaboration of a project that concerns them.

Public participation is an essential element of decision-making, which is necessary to improve its quality and legitimacy. It is a decisive factor in building trust between actors, in particular through its contribution to greater transparency. In order to do so, it requires the accurate resources for its implementation.

The Charter of Public Participation states the values and principles that set the foundation for a virtuous participatory process. It is aimed at all participants – project leader and public – and is an aid in the implementation of the participation scheme. Adherence to the Charter means implementing the values and principles it contains.

The values and principles set out in the Charter cannot replace compliance with existing laws and regulations with which they converge to work to improve the culture of participation.

Figure 4.1. First page of the Public participation charter

Based on the principle that the success of participation depends to a large extent on the degree of trust that the parties place in each other, it addresses all participants (project leaders, associations, citizens) and creates reciprocal engagement. The elements contained in the charter are likely to help create and maintain this trust.

A participation charter is aimed at anyone likely to be concerned by public participation mechanisms: project leaders, both public and private, as well as stakeholders who demonstrate interest in the project (community-based organisations, collectives, citizens, companies, etc.), or organisations that help to make participation a reality and promote it (consultancy firms that provide methodological support to project leaders, civic groups that promote public participation, the CNDP, etc.). It thus highlights that good conditions for dialogue are not the sole responsibility of the project initiator, but also of those who come to discuss it with them. The Public participation charter helps provide favourable conditions for this encounter.

Each stakeholder can apply it to their level; for example, a citizen who adheres to the charter may ask a project leader or their local authority to subscribe to this common frame of reference as a framework for discussion, which takes the form of: "I undertake to apply the values and principles of this charter as the frame of reference for our discussion of project X. Are you willing to make the same commitment for the proper conduct of our debate?"

This charter gives substance to the legal principles set out in the Order of 3 August 2016. This is why both texts were drawn up in the same time-frame and were published almost simultaneously. This tool aims to show a coherent and multi-scalar action of the French government, on the two components "hard law" and "soft law", through their mutual reinforcement.

This tool also aims to contribute to the development of a culture of public participation as an essential element in the construction of sustainable projects (Rio Declaration, Article 7 of the 2004 Environmental Charter).

To date, more than 220 structures and citizens have committed to applying it in their participatory mechanisms.

▸▸ What can we expect from a warrant?

The role of a warrant (whether an individual or a group of individuals) is to ensure the sincerity and smooth running of a consultation. In concrete terms, I see a "firm part" and a "conditional part" in the implementation of this role.

The "firm" side entails ensuring the transparency and completeness of information, making sure that the project leader responds to questions raised by the public. It also means applying the standards for qualitative dialogue that the CNDP has set: independence, neutrality, transparency, equal treatment, argumentation. Further, the warrant can also be seen as a facilitator, or as an advisor on consultation methods, so that they are well adapted. Some stakeholders may even expect mediation.

The listing of these different qualities and abilities highlights how difficult it is to put them all together in one person.

Above all, I see the figure of the warrant as a decisive step forward in bringing the dialogue to a certain level of quality. The warrant is there to guarantee the process itself, and not to take sides on the substance. Paired with the charter, the warrant is a good match for creating a favourable consultation framework.

However, care should be taken to avoid extending the list of what is expected from a warrant. The success or failure of a consultation does not reside with them. They play an important role, but this should not absolve anyone of their responsibilities.

It seems to me that the issue of guaranteeing processes questions, in a much more global way, a constantly increasing need for security and control in public decision-making. This is undoubtedly a corollary of participation: I will get involved if and only if my invested time and energy "serves a purpose". Yes, but what purpose? I cannot go into too much detail here, but let us bear in mind a few obvious points: firstly, we will never get everyone to agree; secondly, the studies and forecasts we make here and now with assumptions in 2020 are unlikely to come true in the end (see Nassim Nicholas Taleb's The Black Swan[2]). We should therefore collectively adopt a modest approach, as there is an inherent element of uncertainty in any project that cannot be deconstructed by studies or by a guarantee mechanism. And given the complexity of the problems that humanity is now facing, even if only considering the management of "common goods", it seems necessary to learn to live with this element of uncertainty, and therefore of risk.

➤ In your opinion, is France ahead or behind other countries in terms of public participation in the environmental sector?

The political-administrative organisations of different countries are so specific and so diverse that I do not think it is possible to objectively elaborate comparison criteria to compare public decision-making systems.

For instance, can environmental management in a federal state really be compared with that of a country like France? Structurally, we are not organised in the same way, the responsibilities of the different levels of decision-making are not the same, and the public decision concerning a railway project, a wind turbine or a public policy linked to the management of water resources probably does not follow the same process in one country as in another.

And this is without taking into account cultural aspects. This became clear during the Covid 19 health crisis; countries around the world are observing and inspiring each other, but with an ability to accept the extremely different constraints, for example between Asia and Europe.

Clearly, comparison is not reason.

It seems to me that the legal framework for public participation in France is very comprehensive and has little to envy others. Are there many countries that have given constitutional value to the principle of participation "in the preparation of public decisions having an impact on the environment" (see Article 7 of the 2004 Environmental Charter), as France has done? To date, whether at the international level (Aarhus Convention) or in our various codes (on the environment, town planning, mining, local authorities, etc.), it seems to me that direct involvement of citizens in the democratic process is today well established and that we can rely on a globally robust system.

---

2. Editions Les Belles Lettres, 608 p., 2012 – Penguin Books, 480 p, 2008

In any case, looking back over the last two years, it is evident that public participation is an exponentially growing dynamic. The Great National Debate and the Citizens' Climate Convention have clearly taken these issues to another level.

Apart from these two very strong democratic experiences, there is a real desire on the part of the French government to shake things up. The Interministerial Directorate for Public Transformation has created a Centre for Citizen Participation, the Ministry of Ecological Transition has its own dedicated participation centre, not to mention the creation of a Ministry for Citizen Participation in July 2020. In addition, some of our operators are embarking on very ambitious experiments. The French Biodiversity Office, for example, has considered that since biodiversity is a common good, which belongs to everyone and therefore to citizens in particular, it is normal to think about how to open up the governance of the Office to citizens and to see what role they want to play in the public policies that it carries out. This seems to me to be a very courageous stance, as it accepts to review its frameworks and ways of administering by giving a significant place to citizens, which is not necessarily self-evident in established systems. But surely, the meaning of democracy is also to provide spaces for citizens to take part in the life in society. I believe that these organisations that dare to question what already exists and what seems obvious are also doing the common good by opening up a path for reflection, as is also the case with the National Food Council, which is working on the link between institutional consultation and consultation with the general public.

In addition to these government initiatives aimed at broadening citizen participation, there has been an extraordinary capacity of local authorities to work towards public participation in the environmental sector for some time now, (regional climate-air-energy plans, etc.) and to invent and reinvent participatory democracy.

▸▸ Why do you think it is important to include citizens in water management?

If there is one thing that is essential to the survival of the human race, it is water! Through the management of this primary resource, a global mindset can be reactivated: understanding that water is not just about turning on a tap, but rather understanding it as a vital and multifaceted cycle, closely dependent on its relationship with its environment, and integrating deep down the fact that it is a common good.

The notion of "common goods" is also frequently encountered when we talk about public participation and what it should mainly be about. Common goods are those resources that belong to everyone and therefore to no one, or the contrary, and which invites us to take a position that stems from a deliberation, a societal choice. We desperately need for citizens' choices to go beyond individual concerns, to always be made with this understanding that we have a common destiny... This, at the State level, is what is called of public interest.

Maslow's pyramid shows that physiological needs require satisfaction first, and even condition the ability to take into account other needs, including security. Yet, the entire water management system goes far beyond physiological needs alone, however in public participation, it is often necessary to "catch" citizens by what concerns them directly, what affects them, their "attachments", as Bruno Latour would say. It makes sense to go back to what is sensitive, to make people understand that a singular need is in fact a question of the survival of the species.

▶▶ With which actors should a culture of public participation be developed and how?

I see developing this culture as a way of revitalising democracy. Public participation complements direct and representative democracy in that it allows citizens to re-enter the public arena on a more continuous basis and with greater power to act. The culture of participation therefore concerns absolutely everyone.

It must be said, however, that this culture is already present, firstly thanks to the regulatory framework which requires project leaders to consult the public, but also thanks to local authorities that have dared to play this hand to the full without the spur of the law.

What is needed at this stage is a change of scale. The demonstrators are there, as are the methods and tools. All that is missing at times is the will.

In order to change scale, the levers that have the greatest power of traction, of suggestion need to be activated. Here, we are obviously talking about education - with more collaborative then competitive teachings, as well as initial and continuous training. But also, and above all, elected officials, who have real powers of transformation, in particular mayors, whose scope for action is more easily identifiable by citizens.

That said, in addition to these great classics—education and elected officials—, I believe that other postures should also be reexamined. I am thinking of the citizens, who sometimes do not realise that administering a country is an infinitely heavy and complex thing. An example would be the Yellow Vests movement[3] which, when it reached a certain critical mass, considered the question of its structure. Who represents the movement? One or more? Appointed or elected? A federal or pyramidal organisation? Who decides what and how? If the Great Debate[4] had one virtue, it was that by confronting themselves with the challenges of democracy, some citizens realised the interest of the institutions already in place. When I say that, I am not saying that these institutions are functioning at their best, as it is obviously increasingly complicated to obtain the assent of citizens to public decisions. However, caution must be taken not to dismiss everything with the sweep of a hand, because our institutions are the result of the long process of democracy.

I am also thinking of the world of research. I am frequently surprised that the academic world, which urges project owners to change their positions, has only marginally found a way to change its own, having only too rarely offered to help shed light on the operational issues raised by major democratic issues. Democracy is being shaken from all sides, some even say it is in danger. So why can't we get the world of research to collaborate with the world of project management in the broadest sense, in order to find the most effective ways of developing a project or a public policy?

---

3. The Yellow Vests movement is a protest movement that began in October 2018 in France to protest against rising motor fuel prices. The name of the movement comes from the fact that many demonstrators wore yellow high-visibility vests.

4. The Great Debate is a french national public debate that was held in France between january and march 2019 following the Yellow Vest movement. Each french citizen could give his/her opinion about four topics (ecological transition, taxation, democracy and citizenship, organisation of the State and public services) through lists of grievances, exchanges between citizens and mayors, local debates, a website, themed national conferences and regional citizens' conferences.

The critical analysis they can produce is, in my view, largely underused when it remains confined to publications that project leaders do not have time to read. Why is it so hard to sit down at the table with the project leaders to inform their thinking with academic elements that allow for central issues to be reviewed when implementing participation in an effective way? Is legitimacy the result of numbers and/or a random draw? (and therefore, as a project leader, do I invite a large number of people or do I choose to use mini-groups?) What do the social sciences have to say? There is no definitive answer, of course, only arguments in favour of one thesis or the other. But helping, for example, to construct "states of controversy" on major democratic issues such as legitimacy, the effects of a guarantee system, the synthesis of contributions (by hand/artificial intelligence) or others, would help to shed light on what is at stake in the public debate and to make sense of it. I therefore have the greatest respect for those who dare to engage in action research, which is undoubtedly an interesting lever for developing the culture of participation.

Beyond the actors themselves, developing a culture of participation must be based on reference frameworks, which provide the opportunity for coherence and standards. The charter is one of them, but I won't go back over it.

Beware, however, of democratic fatigue, born as much from the multiplication of requests as from discouragement when the link to the decision is not sufficiently evident.

Finally, I would say that beyond the legal texts and reference frameworks, beyond the methods and tools, public participation is above all a form of spirit, an attitude rather than know-how. It is when each person embodies it in their daily life, in their relationship with others, that it takes root.

Territorial facilitator, a profession to be developed and defended: A Tunisian experience

Houssem Braïki, Guillaume Lestrelin, Sylvie Morardet, Soumaya Younsi, Emeline Hassenforder, Amar Imache, Audrey Barbe, Anissa Ben Hassine, Fethi Hadaji and Mohamed Chamseddine Harrabi

Territorial facilitators work towards facilitating the dialogue between a wide range of stakeholders (farmers, elected officials, administrators, etc.) with regard to developing a territory. This chapter describes a pilot project in Tunisia in which agents from regional agricultural services were trained and accompanied in implementing their new profession for concerted territorial planning in six rural areas of Tunisia.

The emergence of the concept of participation—in the sense of contribution by citizens to political processes and decisions—in the public sphere dates back to the mid-twentieth century. As a major demand of civil society, within the broader social movement and fight against inequalities of the 1960s and 1970s (Wuhl, 2008), the concept of participation was gradually formalised, institutionalised and integrated at the international level, for example, as a fundamental principle of sustainable development (see the Rio Declaration in 1992) and translated into legislation at the national level (see the French law of 2002 on local democracy). In practice, participatory approaches are now being implemented throughout the world, at various scales and in a multitude of areas (e.g. participatory management mechanisms in companies, participatory budgets in municipalities and regions and, to a lesser extent, citizens' conventions in support of governments).

Spatial planning and natural resource management have not escaped this trend. In areas of public intervention that have to deal with a diversity of actors and interests, faced with issues of social justice and conflict management, participation appears to be a means of making more consensual, and even fairer, decisions concerning local development orientations and strategies, rights and rules for resource use, etc. However, participation cannot just be decreed; it must be "fitted out" with sociotechnical mechanisms (institutions, operating rules, decision-making processes, etc.) that allow for interactions between actors to be organised and which make participatory processes legible. And, in many circumstances, these mechanisms require facilitation. Their implementation and dynamics often depend on the intervention of facilitators in charge of getting the actors to interact, developing or maintaining collective action, without however influencing decisions (Dionnet *et al.*, 2017).

By including participatory democracy as a fundamental principle in its Constitution in 2014, the public authorities of post-revolutionary Tunisia have embarked on this major project. Recently, several ministries involved in land use planning and natural resource management have adopted strategies to promote citizen participation. The national strategy for the development and conservation of agricultural land (2017), in particular, recommends the implementation of a participatory approach (consultation and facilitation) for all rural development projects. It is within this framework that, since 2018, the Climate Change Adaptation Programme for Vulnerable Rural Territories (PACTE[1]) has been supporting and training territorial facilitators, called Rural Development Officers (RDO), who are in charge of mobilising various local and regional actors (local communities, municipal councils, civil society, administrations, private sector, etc.) within a territorial planning mechanism.

▸▸ The usefulness of training and the role of facilitators in practice: testimonies from the Rural Development Officers

The RDOs themselves testify to the evolution of their skills and their posture. According to them, the acquisition of theoretical knowledge or *"new scientific notions"*, and know-how was facilitated by the frequency of the training sessions (*"Every two weeks, we have the chance to meet together to exchange, discuss and learn together. We have gained a lot of knowledge about group facilitation techniques, it is new knowledge and a new experience"*), as well as their adaptation to the pace of the programme (figure 5.1).

Figure 5.1. Training sessions strengthen the capacities of the facilitators, rural development officers

Another aspect of the training scheme considered important by the RDOs concerns the diversity of participants in the sessions. These sessions *"brought together administrative agents from different 1) specialties, 2) areas and 3) topics, all united to be trained as facilitators for spatial planning"*, as well as researchers from different disciplines. The RDOs

1. PACTE = Programme d'adaptation au changement climatique des territoires ruraux vulnérables

also pointed out the importance of the diversity of shared viewpoints and the originality of the training session set-up (alternating theoretical and practical sessions), as well as the richness and relevance of the interventions. According to them, this sharing of skills and experiences between agents from different specialties (soil, water, forestry, water and soil conservation, etc.) was a great added value in the learning process.

Finally, a study tour to France was cited as a highlight of the training process: *"The study tour to Montpellier, to observe experiences in a different territory and discuss with farmers and researchers from abroad, greatly helped to advance our skills. It is all about discussing with other actors and farmers and learning about other methods and practices. During my discussions with farmers in Bizerte, I can draw on this experience abroad to share targeted and beneficial knowledge".*

The RDOs mentioned several situations in which they were able to mobilise the skills they had acquired:

– An RDO from Bizerte recounts negotiations with a farmer. The aim was to get the farmer to give up a 400 m^2 plot of land for free so that a collective borehole could be dug on the territory: *"I was able to negotiate with a farmer for the common good of the zone by using facilitation techniques and constructive discussion, without being nervous or shy and without the fear of doing something wrong; all this thanks to the comments of the trainers and researchers who accompanied us during the training sessions and the simulations we carried out together during these sessions".*

– An RDO from Kairouan mentions the ability to facilitate discussions between high-level actors: *"Today, we facilitated a discussion on the territory committee between elected officials, the mayor and the department head at the RCAD*[2]*. And having followed the fundamentals of facilitation techniques, this meeting was a success. In addition, we also drew up quality minutes written in French".*

– Another RDO explains that their French writing skills have improved since writing over ten diagnostic reports on small local territories as well as a summary report: *"We have improved our writing as well as our speeches in French following 1) the transcription done during diagnostic 2) the constructive comments in the Word files of reports 3) the close support of the referent researchers".*

Finally, the RDOs evaluated their own evolution in terms of interpersonal skills over time: *"After participating in PACTE training courses on consultation, conflict management, systemic participatory diagnostic, etc., a change in posture and reaction was noticed".* They emphasise the importance of honesty and sincerity in facilitating a discussion: *"I am very honest and spontaneous with people, especially farmers. You have to be clear with the locals if you can't do anything on their land. Sharing accurate information with them is important in creating trust".*

▸▸ From land planner to territorial facilitator: connecting theory to practice

The work of the territorial facilitators has enabled significant participation of the local population of the intervention zones, in both territorial diagnostic and development plan elaboration. This participation goes far beyond what is usually observed in development programmes of this type in Tunisia (Burte *et al.,* 2017).

---

2. Regional Commission for Agricultural Development

During the participatory diagnostic phase, over 4,100 residents (out of a total population of approximately 26,000 in 2014 in the six intervention zones) participated in the events organised by the PACTE programme (figure 5.2). Particular attention was paid to the participation of women, who represented about a third of the participants, which is not very common for work in rural Tunisia.

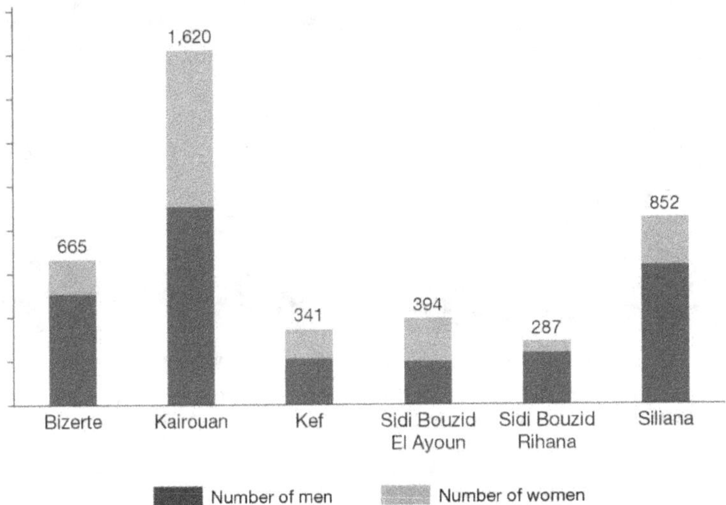

Figure 5.2. Number of participants in the diagnostic feedback sessions organised by the PACTE programme (Total = 4,159)

A series of workshops was organised to present and discuss the diagnostic with the local population (figure 5.2 and 5.3). On this occasion, the participants voted on the development issues they considered most important for their territory (figure 5.4).

During these workshops, the participants indicated what actions they wanted implemented to meet these challenges. Proposals for action were also collected from those who had not been able to participate in the workshops, through forms left with contacts in each territory. In total, around 11,400 proposals for action were collected and entered into a database. This database was used at a later stage by experts and committees representing the local stakeholders as part of the territorial planning framework.

The involvement of territorial facilitators in the programme's approach has profoundly changed their professional practices and, in particular, their attitude towards the local population, as revealed by interviews conducted with them (see previous section) and their work colleagues (Jendoubi *et al.*, 2021). The facilitators are more attentive to the locals than before; they ensure that public policies are consistent with the issues identified with those living in these zones; and they often position themselves as the locals' spokespersons to their colleagues, pending the appointment of territorial representatives.

The impact of the participatory process undertaken within the PACTE framework on the practices of the public servants goes beyond the territorial facilitators. Indeed, throughout the process, the facilitators involved other agents from different departments of the RCAD and beyond; they called upon the knowledge of their colleagues in the programme's target territories to establish contacts with local actors in the

Figure 5.3. Feedback from a diagnostic of the Kairouan action zone

During the diagnostic phase, participatory workshops were organised to report and discuss the results obtained. This initiative aims to strengthen the validity and reliability of the data collected during the diagnostic phase.

Figure 5.4. Voting on territorial issues in the Kairouan zone

The participants engaged in a process of selecting the territorial challenges by means of a vote. This participatory approach enabled the participants to prioritise and determine the issues they considered to be the most crucial for their area.

diagnostic preparation phase. They also involved them in carrying out individual and group interviews with local residents during the diagnostic phase and in facilitating feedback workshops with the population. In addition, exchanges between the different departments of the RCAD and with departments of other sectors (e.g. health, education, environment, equipment…) took place during the meetings of the programme's Operational Monitoring Committees in each intervention zone. Finally, experts from different fields were called upon to discuss and supplement the proposals for action made by members of local communities.

This wider involvement of the public administration in the territorial planning approach has enabled the territorial facilitators to transfer tools and methods acquired during their training to other colleagues, some of which (such as "participatory maps"[3]) they are ready to use in other circumstances.

The emergence of the territorial facilitator, a new profession, has also led to changes in the professional network of the agents involved. Indeed, connections within the teams of facilitators have been strengthened in each intervention zone as a result of working together almost daily on the PACTE programme. New connections were also created between facilitators from different zones thanks to the training courses they attended together and the sharing of experiences between sessions. Other connections were created between facilitators and agents of the agricultural administration at different levels:
– Local (agents from the Territorial extension units),
– Regional (agents from other RCAD departments, the Agricultural Land Agency or the Livestock and Grazing Office),
– National (agents from the General Directorate for the Development and Conservation of Agricultural Land).

In the end, the professional network of territorial facilitators has been extended to other administrative sectors outside the agricultural sphere, in line with the PACTE programme's objective of integrated and multisectoral planning (Jendoubi *et al.*, 2021).

Finally, the perception of many agents of Tunisian agricultural administration has evolved thanks to the emergence of this new role of territorial facilitator. Most of the colleagues involved in the participatory approach appreciated it. They emphasised the trust created between the facilitators and the local population and the in-depth knowledge of the territories provided by the local population, which will help facilitate the implementation of the actions and the sustainability of PACTE programme investments. However, some believe that it will be difficult to generalise this approach without more fundamental changes in the way the Tunisian administration intervenes in order to reduce procedural constraints, time and cost of implementation.

▸▸ Conclusion

Facilitating a participatory approach to territorial development cannot be improvised. It requires the acquisition of theoretical knowledge, know-how and interpersonal skills. A training cycle combining theoretical knowledge, simulations, practical application in the field and shared reflective analysis between participants was designed

---

3 This tool was used to help farmers delimit their territory, highlight the distribution of key resources (water, vegetation, relief, etc.) and share their understanding of its current situation and evolution.

and implemented for two years in Tunisia as part of the PACTE programme. This experience represents a first step in the creation of the new profession of territorial facilitator, an essential link in the implementation of a concerted territorial planning process in rural areas. The continuation and extension of this experience to all rural areas of Tunisia faces a number of challenges.

The first challenge is to formalise and simplify the training curricula of future territorial facilitators without altering what makes it rich. These training courses should encourage the construction of composite expertise that integrates, beyond conventional technical knowledge, knowledge of human and social sciences (i.e. principles of the participatory approach and integrated rural development, consultation engineering, communication, and conflict management) and an apprenticeship in appropriate postures. The second challenge is to create favourable conditions for networking, which helps to put into practice the theoretical knowledge and skills acquired in training through exchanges of experience with peers, experts and scientists. The third challenge consists in raising awareness (or even training) facilitators' colleagues and hierarchy, as well as the elected officials of the territories where they work, in the principles of the participatory approach so that they can contribute together to the co-construction of action programmes in consultation with the population and to the development of ad hoc public policies. The fourth challenge is to formally recognise this new profession within the Tunisian public administration, which implies defining the required levels of competence and the specific position of the profession in the structural hierarchy. This also raises the issue of RDOs' affiliation with the agricultural administration (RCAD and Territorial extension units). The fifth and final challenge is to provide the logistical means for facilitators to carry out their activities.

These challenges are not only present in Tunisia; they are in fact favourable conditions for the implementation of territorial facilitation, whatever the territory. Ultimately, the institutionalisation and formalisation of the profession raises the question of its generalisation, as it is a profession which, by its very nature, relies on the personal and creative capacities of the individuals who embody it. This implies that even with an equivalent level of training, certain profiles would not be suitable for this position.

▸▸ References

Burte J., Chevrillon A., Ben Haha N., 2017. Vers une territorialisation des politiques rurales en Tunisie : l'exemple des politiques de conservation des eaux et des sols. *In :* Caron P. (éd.), *Des territoires vivants pour transformer le monde*, Versailles, Éditions Quæ, 172-178.

Dionnet M., Imache A., Leteurtre E., Rougier J.-E., Dolinska A., 2017. *Guide de concertation territoriale et de facilitation*, Lisode : Montpellier. 64p.

Jendoubi M., Hassenforder E., Lestrelin G., Imache A., Braiki H., Barbe A., 2021. Apprendre la participation au contact des facilitateurs ? Partages de compétences et de posture au sein de l'administration agricole en Tunisie. *Alternatives Rurales*, 8:90-107.

Wuhl S., 2008. *La démocratie participative en France : repères historiques.* Institut de recherche et débat sur la gouvernance, Programme coproduction du bien public et partenariats multi-acteurs. http://www.institut-gouvernance.org/fr/analyse/fiche-analyse-418.html

Chapter 6

Participation and the construction of social acceptability: Fantasy or reality?

Benjamin Noury and Laura Seguin

This chapter reviews the origins of the notion of social acceptability and then proceeds to describe the participatory strategies associated with it. Some of these strategies are more akin to manipulation than an aspiration for co-construction. In this chapter, two case studies on the implementation of technological innovations demonstrate how this pitfall may be avoided. For each case, the approach provided insight into an issue of public concern, leading to an arena of choice in which several technical solutions were open for discussion.

⯈ From environmental disasters to the construction of social acceptability

Social acceptability is on the rise in this time of Covid! Since the beginning of the pandemic, the scientific council and the government of France have been thinking about how to get citizens to accept measures to combat the spread of the virus. From the outset, scientific knowledge and expertise have been mobilised to build this acceptability.

Yet, since the 1986 Chernobyl accident and the minimisation of the risks associated with the radioactive cloud, the legitimacy of scientists to "get people to accept" public action has taken a hit. Social acceptability is a concept that has resulted from the increase of exactly this type of environmental or sanitary disaster. In the 1960s, scientists and environmental movements questioned the values of technical progress and were concerned about the environmental impact of major developments. These initial concerns were then confirmed by a series of major industrial incidents: sinking of supertankers (Torrey Canyon in 1967; Amoco-Cadiz in 1978), toxic fumes (Seveso in 1976), hydrocarbon explosions (Los Alfaques in 1978), nuclear accidents (Three Mile Island in 1979; Chernobyl in 1986), as well as explosions and chemical pollution (Bhopal in 1984; Sandoz in 1984).

The increasing frequency and magnitude of these disasters is leading to a growing number of citizen actions and engagement. The development of large-scale projects involving the exploitation and use of natural resources is being contested and social acceptance is becoming a key issue for the promoters of these projects. The latter are putting in place strategies to avoid actions and mobilisation that could slow down or block the implementation of projects. These social acceptability mechanisms aim to establish trust and enlist the conviction of elected representatives and populations.

If a project is well explained, in a form of technical rationality, it will be accepted. Similarly, once a project is in place, permanent mechanisms such as local information and monitoring committees (CLI or CLIS[1]) are today compulsory for facilities classified as environmental protection.

Mobilisation and action strategies take into account the adhesion or project support strategies deployed. Rallies and other movements are most often not limited to the projects themselves. Their scope also questions the vision of the world that led to the conception of these projects. Over the last years, space for public debate has been defined so the various parties may express themselves and confront their positions with the aim of enlightening the decision-makers in their choices. The French National Commission for Public Debate, for instance, is supposed to ensure the proper implementation of such consultations. These arenas are designed to facilitate cooperation between "experts" and "lay people". The aim is to engage partners in collective dynamics where the emphasis is on "stakeholders" instead of "audiences". Strategies to gain project support then come into play, which at times near what could be seen as manipulation of the debate mechanism. Certain public debates have consequently been the subject of much criticism, particularly from stakeholders who denounce the manipulation of "ordinary" citizens through this type of mechanism (box 6.1).

Box 6.1. Two cases of public debate under pressure

In 1998, the first citizens' conference organised in France dealt with the controversial issue of GMOs. It was led by the Parliamentary Office for the Evaluation of Scientific and Technological Choices. The feedback from this experience was rather mixed. The researchers who followed the process emphasised the quality of the report, in which arguments were particularly well set out and detailed. They however noted that this debate took place in a public arena where considerable militant mobilisation was already engaged (environmental associations, agricultural unions), which severely denounced the debate as an attempt to manipulate public opinion (Joly and Marris, 2003).

In 2009, the debate on nanotechnologies organised by the National Commission for Public Debate was highly disrupted by opponents of this technology. They denounced an attempt at instrumentalising the debate by resorting to "lay" citizens, with the aim of bypassing any protest and challenges, and thus reserving the right to define this technology as a responsible innovation (Laurent, 2010). Faced with what they considered to be an operation to manipulate opinion in order to gain social acceptance of nanotechnologies, the members of the *Pièces et Main d'œuvre* collective in Grenoble actively demonstrated their opposition to the holding of a public debate by disrupting its progress*.

* A leaflet distributed on this occasion is available on their website: https://www.piecesetmain doeuvre.com/spip.php?page=resume&id_article=230

▸▸ Getting people to accept by getting them involved

Beyond major national debates, participatory approaches are also being developed at the local level for the social acceptability of technological innovations. This is particularly the case for research projects with an operational aim that are carried out in

---

1. CLI or CLIS = commissions locales d'information et de suivi

partnership with socio-professional actors. The consent obtained from the public at large via a participatory process thus appears as a "sugar coating" for project approval, a type of acceptology. These types of debate are generally limited to the study of a given solution and any possible adjustments. The initial issues and upstream decision-making processes that led to the solution are, however, not addressed. Alongside scientists and engineers from the natural sciences, social scientists are called upon for their skills in "social" and/or "participatory engineering", leading to a certain awkwardness (Barbier and Nadaï, 2015).

We share this discomfort with regard to the notion of social acceptability; however, we find it necessary to adopt the term used by operational actors so as to start from a common base language. The intervention of social science researchers should, nonetheless, provide a reflexive view of this notion and the use of participation in this context. Indeed, the risk of resorting to participation to "get people to accept" the choices made by decision-makers and experts is first and foremost a risk that concerns the quality, and even the success, of the dialogue initiated with the stakeholders who are to be mobilised/convinced: "When the debate is designed as a means of revealing a truth, and not as the exploration of different scenarios, it produces as many conflicts as it resolves" (Barthe, 2005).

In the two cases we describe below, a participatory approach was initially set up to respond to a problem of acceptability. They however turned into debates that highlighted a public concern and thus opened up a space in which various technical solutions could be discussed.

▸▸ Case no.1: Water reuse

In the context of climate change, water reuse is a frequently cited complementary resource as an alternative to water abstraction. It consists of recovering wastewater from different sources, treating it to remove impurities and then using the water again for other purposes.

This practice is still underdeveloped in France compared to other Mediterranean countries. In 2016, the Rhône-Mediterranean-Corsica water agency launched a call for projects on the subject to make up for this delay in France. The *Read'Apt* project, winner of this call, aims to assess the relevance and feasibility of reusing treated wastewater in the Luberon region, which suffers from repeated severe droughts. An experimental plot was set up to test and analyse the quality of the water, soil and irrigated plants (figure 6.1). A study evaluated the potential of different sites in the area and the final part of the project focused on the social acceptability of water reuse.

A series of interviews and surveys were carried out in this area to identify the perceptions associated with water reuse. The practice is largely unknown. It is often associated with grey water recovery on a domestic scale. It does not generate a visceral reaction of rejection but raises questions about health risks.

Subsequently, several activities were tested to highlight this practice and debate its relevance within the territory. The aim was not to persuade or get people to accept the practice in a covert manner but rather to bring this alternative to the fore as a choice. A website and a public meeting on water reuse were proposed. These activities attracted a very limited audience despite extensive communication. It is important to note that, for the time being, implementation of water reuse remains very hypothetical in this territory.

Figure 6.1. Experimental site for water reuse, Saint-Martin-de-Castillon (Vaucluse)

This limited turn-out led to a change of approach by:
– more explicitly addressing the challenge to be tackled: dealing with drought,
– joining forces with local and non-institutional community-based organisations.

To this end, a game workshop was set-up at local events that focused on the environment, agriculture and food. The deployed game scenario, developed according to Wat-A-Game principles (chapter 13), features mayors and farmers who face a decrease in the availability of water from the river they share (figure 6.2). Water reuse is one of the options available to the players to cope with this water shortage.

These sessions brought forth many questions on water management and governance as well as on how sanitation functions. The debates focused on divergent views concerning the management of supply (increasing the volume of available water) and demand (implementing water saving measures). Water reuse was at the heart of these discussions, presented not as *the* solution but as one option among others.

▸▸ Case no.2: Artificial wetland buffer zones

Artificial wetland buffer zones (AWBZ) are an ecological engineering solution, or "nature-based solution". These are artificially recreated ponds that allow for the recovery of agricultural water and the "natural" purification of some of the contaminants (nitrates, phytosanitary products) before it filters down into the ground to the water table.

In 2010, in the scope of an experimental project in the Brie region of Seine-et-Marne, Irstea (INRAE since 2020) researchers, who came up with this innovation, and local partners initiated an AWBZ development project comprising 13 pre-identified sites on agricultural plots (figure 6.3). Dialogue with the farmers took place after the diagnostic and project definition stages and was in the form of a negotiation. The aim was to implement a technical innovation that was pre-constructed by experts on pre-identified sites. The scientific rationales, particularly hydrological, that led to the design of the works, were confronted with many other rationalities, the farmers' financial and

Figure 6.2. Eau'Sec role-playing workshop in Forcalquier (Alpes de Haute-Provence)

operational concerns (the loss of productive agricultural land, the maintenance and monitoring of future facilities, the risk of proliferation of invasive species), as well as more symbolic and political issues related to the visibility of the problem of diffuse pollution brought to light through the proposed facilities.

In 2017, the researchers and territorial stakeholders involved in this first experiment put the issue of territorial dialogue at the heart of a new research project entitled *Brie'eau*. In collaboration with social science researchers, a completely different dialogue concept was put forth:
– a debate not limited to the curative solution represented by water reuse, but which posed the problem of diffuse agricultural pollution more broadly, and opened up the range of possible solutions (in particular changes in agricultural practices);

Figure 6.3. Artificial wetland buffer zone (AWBZ) in a crop farming context, Seine-et-Marne

– a circle of participants not limited to the farmers directly concerned by the technical solutions, but which included the diversity of local stakeholders concerned by the public problem under discussion: elected officials in charge of drinking water, river unions, stakeholders from the agricultural sector, user associations (hunters, fishermen), environmental organisations.

The participatory approach was implemented over a year, combining meetings in the field, workshops for participants to exchange their views and representations, workshops to simulate agronomic scenarios for the territory (integrating changes in practices and buffer zone-type landscaping), and a role-playing game as a virtual space for discussion and negotiation around individual and collective actions (figure 6.4).

Over time, it became apparent that the focus of the dialogue shifted from a technical solution deemed optimal to one of public concern, that of diffuse agricultural pollution. This shift highlighted possibilities that became apparent during the process. Far from being restricted to technical or ecological innovations primarily concerning the farmers, discussions also touched on organisational, economic and social innovations, with a desire for closer collaboration between local stakeholders from compartmentalised worlds (agricultural stakeholders, water stakeholders, citizen-consumers).

▸▸ Conclusion

Technological innovations bring with them technical, regulatory, economic and social uncertainties. Participation can be mobilised as a tool to facilitate the reception of these innovations in society. Indeed, participation provides a space for mediation to take place and thus allows for technical and social issues to come together.

Figure 6.4. Test run of the Rés'eaulution Diffuse role-playing game

However, one must not take precedence over the other. Transformations must be a two-sided keeping the spirit of integration and co-construction in mind. Even in times of crisis, authoritarian orders to get measures accepted seem to have their limits. The anti-mask movements and calls from the medical world to return to a democratic health system, which has been eluded since the beginning of the Covid 19 pandemic, attest to this (Legros, 2020).

▸▸ References

Barbier R., Nadaï A., 2015. Acceptabilité sociale : partager l'embarras, *VertigO – La revue électronique en sciences de l'environnement*, 15 (3). https://doi.org/10.4000/vertigo.16686

Barthe Y., 2005. Discuter des choix techniques, *Revue Projet*, 284, 80-84. https://doi.org/10.3917/pro.284.0080

Joly P.-B., Marris C., 2003. La participation contre la mobilisation ? Une analyse comparée du débat sur les OGM en France et au Royaume-Uni, *Revue internationale de politique comparée*, 10, 195-206. https://doi.org/10.3917/ripc.102.0195

Laurent B., 2010. *Les politiques des nanotechnologies : pour un traitement démocratique d'une science émergente*, Paris, Charles Léopold Mayer, 242 p. https://doi.org/10.3917/pox.096.0187

Legros C., 2020. La démocratie en santé, victime oubliée du Covid-19, *Le Monde*, https://www.lemonde.fr/idees/article/2020/09/25/la-democratie-en-sante-victime-oubliee-du-covid-19_6053545_3232.html (consulté le 3/10/2020).

Chapter 7

Supporting participatory processes in territorial governance: The researcher's "risky" stance

Testimonials from Brazil, Tunisia and New Caledonia

Caroline Lejars, Veronica Mitroi, Guillaume Lestrelin,
Julien Burte, Isabelle Tritsch and Nils Ferrand

Based on various testimonials from researchers involved in accompanying large-scale participatory and transformative projects, this chapter identifies and discusses some "risky stances" and frictions that researchers may encounter, as well as the strategies they develop to cope with them. The chapter shows that the researcher's stance, understood as his/her personal positioning in terms of theoretical and methodological choices and interpersonal interactions with other stakeholders, is a key element in the dynamics of the participatory process, even though it is very often neglected. The chapter brings valuable contributions for developing the reflexivity of researchers and project managers regarding their own role in transformative participatory processes.

Setting up participatory research for natural resource management is not a neutral act, particularly when it aims at democratisation and/or local governance (Crémin *et al.*, 2018). The role researchers play in this process is worth noting and may be an issue since, as D'Aquino and Seck Sidi (2001) point out, "every development programme brings its own implicit political ideology". This is true both for development research projects (Olivier De Sardan, 2022) as well as for participatory research projects when embarking on strategic planning, development schemes, governance mechanisms or water policies.

As researchers involved in these participatory research projects, we provide guidance and support to organisations or individuals who are locally involved in decision or change processes. This "support" is an integral part of action research projects, can take a wide variety of forms beyond the mere production of knowledge and may include providing advice, developing methodologies, leading the process itself, and so forth. For researchers, being a stakeholder in these transformative participatory projects often involves negotiating both what is expected of them from the other players (funders, project partners and scientific managers) and their own position in the research and support system (Daré and Venot, 2016; Barnaud *et al.*, 2016). Researchers may find themselves torn between their epistemological and ethical research framework, their scientific objectives and the very diverse expectations of the stakeholders with whom they work (funders, state and political stakeholders, managers and citizens, etc.).

While the analysis of participatory processes at work in natural resource management can fuel critical reflection on existing frameworks for public action, it can also be particularly difficult to maintain this same analytical and critical status when involved in a participatory research project aiming at transformation. The epistemologies and ethics of management sciences (David *et al.*, 2012), intervention research (Buono *et al.*, 2018) and even intervention sociologies (Herreros, 2009) have largely shown that researchers involved in action research or transformative processes cannot maintain the axiological neutrality[1] (Weber, 1965) characteristic of descriptive analytical social sciences. They are themselves stakeholders in the process, sometimes even the initiators of a change project, and work with other actors to transform the social world, not just observe or describe it. This research stance is thus not neutral, and all the more so in participatory research projects that are strongly shaped by public funds for development and/or driven by democratisation and/or transformation of resource governance.

We define the stance of the researcher who is supporting participatory schemes as his/her personal positioning, expressed through his/her theoretical and methodological choices (Charmillot, 2021), inter-personal relationships and "alliances" that she/he manages to build (Akrich *et al.*, 2006), or the way in which she/he perceives knowledge and the ways in which this knowledge is constructed and shared (Mazzocchetti, 2007). We consider that this stance is likely to evolve over time, including during the course of a research project (Ballon *et al.*, 2019; Brun *et al.*, 2007), and that it is linked to an inter-relational dimension that takes into account the position that the researcher occupies in relation to his/her research objects, interlocutors, field, as well as his/her peers and the institutions that structure and/or fund his/her activities (Alphandéry and Bobbé, 2014).

This chapter aims to present the dynamics of the researcher's stance when involved in a participatory process that supports governance and to discuss the difficulties and frictions she/he may encounter. We take a reflexive approach to analysing the case of researchers involved in large-scale participatory processes (large population, transformative impact) to co-construct regional governance schemes on the topic of water or associated public policies. Our focus is mainly on the field of water democracy through the exploration of the researcher's position within participatory processes that have an impact on institutional decision-making. This chapter does not aim to provide an exhaustive analysis of possible stances. It aims instead to show, using a few examples and critical accounts, how the diversity of implicit or explicit expectations and objectives of the participants, project sponsors and researchers can generate biases and changes in a researcher's stance over the course of a project, and beyond that, how this stance can then be taken into account. We elucidate these biases in order to highlight the tensions they can generate in terms of research stances, the responses provided and the ensued learning. Building a scientific stance while being involved oneself is often a matter of individual trial and error, and we feel it is necessary to create forums for sharing experiences.

We have taken three large-scale citizen participation projects, in Tunisia (see chapter 5), New Caledonia (see chapter 18) and Brazil as examples. These projects have in common:

---

1. Axiological neutrality, or science that is free of value judgements, as theorised by Max Weber (1917), is a methodological stance adopted by the researcher who attempts to become aware of his own values in order to reduce in as much as possible the bias that his own value judgements might cause in the research at hand and in the interpretation of the results.

– the fact that they were initiated and are supported by governments;
– the desire for strong citizen involvement, in contexts where participation is not a given;
– processes which generate proposals that can challenge the authorities in place, authorities with whom we work.

We begin by describing the specific features of the processes that have been implemented. Using testimonials and feedback from the researchers involved in these projects, we then discuss the different roles or expectations perceived by the participants with regard to the researchers, and the objectives and expectations specific to the researchers, in order to elucidate the stances and roles of the researchers during the process, as well as any chosen or imposed consequences. Finally, based on collected testimonials, we illustrate how involvement in these projects can constitute a "risky stance" for researchers. We conclude by stressing the need for researchers to reflect on their actions, and the need to develop tools for clarifying objectives and preparing consultation, as well as ways of guaranteeing and monitoring the role of researchers.

▸▸ Case studies: participatory processes inspired by CoOPLAGE, strongly linked to political decision-makers, funders and citizens

The three participatory projects given as examples and carried out in the South by the authors of this chapter[2] focus on transformative processes such as the development of public policy (policy guidelines, action plans) and support for the development of local and decentralised systems of governance. The CoOPLAGE approach (see chapter 2) was used in a different way for each of these projects.

Presentation of the case studies

These projects, which are at various stages of completion, aim to:
– improve the involvement of local stakeholders and citizens in policy-making on water and other natural resources;
– promote shared diagnoses and action plans at the local level;
– highlight the attractiveness that such involvement can have for decision-makers at the regional and national level.

The underlying assumption of all these projects is that the participation of citizens and local stakeholders in the development of public policies improves efficiency, and facilitates the implementation of sustainable actionable solutions and functional governance systems.

The Sertoes project in Brazil for the sustainability and hydric resilience of north-eastern territories

The Sertoes project is a research-action project designed to help engage stakeholders in the co-construction of a multi-level territorial water governance model, using an innovative multi-actor process designed and led by the researchers themselves. This process is fuelled on the one hand by the production of knowledge on the state of water resources and their territorial uses, and on the other hand by the strengthening

---

2. Caroline Lejars and Nils Ferrand took part in the project in New Caledonia; Guillaume Lestrelin, Julien Burte and Nils Ferrand in the project in Tunisia; Veronica Mitroi, Isabelle Tritsch and Julien Burte at various stages of the project in Brazil.

of stakeholders' capacities. The project management approach is based on the implementation of an iterative and progressive process, which involves both the production of cross-sectoral and multi-scale expertise on water resources, and the empowerment and support of stakeholders in territorial planning and water resource management processes. This process allows for information to be produced, learning to take place and stakeholders to be mobilised around the three phases of the project:
— diagnostic or analysis of past/present trajectories,
— co-construction of sustainable and resilient future trajectories,
— and construction of a methodology and implementation of a pilot linking a Territorial Intelligence System (TIS) and a territorial water governance system.

The project was launched in March 2021 for a period of three years, and at the time of writing was still in the defining and testing phase of the local governance model.

The PACTE programme in Tunisia to help vulnerable areas adapt to climate change

Tunisia's Programme for Adaptation to Climate Change in Vulnerable Territories (PACTE) was launched in the wake of the Arab Spring, at a time of democratic transition and the strengthening of the role of local authorities (Dafflon and Gilbert, 2018). It was initiated and implemented by the Ministry of Agriculture. This programme aims to plan and finance actions that promote the sustainable management of natural resources, the development of the agricultural and forestry sectors and the strengthening of local governance in six rural areas. A group of fifteen-some French and Tunisian researchers is involved in guiding this programme though a sequence of projects spanning more than eight years, leading a co-design process for structuring multi-stakeholder platforms for territorial diagnostic, citizen debate on development issues, concerted planning and monitoring-evaluation of investments and their impact. These platforms should in particular act as supports for large-scale local participation, in a context where there are few or no intermediary organisations that are functional and recognised as legitimate by local communities. The programme was launched in 2018, following two successive small-scale projects (in 2012 and then 2014) which helped to co-design a methodological pilot for the territorial diagnostic and planning approach. The PACTE programme is now in its final stage (i.e. investment and implementation), which will run until 2027.

Supporting the construction of a Shared Water Policy in New Caledonia

A team of researchers provided methodological support over three years in the form of a project cluster, initially to train stakeholders at the request of the Northern Province (2016 to 2018) and then, at the request of the New Caledonian government's services (2018 to 2019), to provide methodological support for the co-development of the country's water policy and, finally, to evaluate the process. Support in the development of the policy is detailed in chapter 18. A shared diagnostic process was set up for all the issues and target sectors, under the guidance of an interdepartmental government group, followed by a broad participatory planning process, which was finally prioritised and broken down into action plans published in a framework document. The researchers proposed the methods and materials, trained some facilitators, supported the process and assessed its progress. The process took place in a politically tense context in which referendums on independence were ongoing.

The researchers' contribution to these case studies

For these three processes, the researchers' contribution was organised around six major components of the participation process, although they were not necessarily sequential, and sometimes not anticipated or planned for at the start of the projects:
– Project initiation and set-up. In all three cases, the project was initiated at the request of local government players. The researchers were more or less involved in the initial framing of the projects, with strong involvement in the Brazilian and Tunisian cases, where the researchers steered the process; involvement in New Caledonia was very marginal;
– Support for the co-designing process of the participatory approach to be implemented. In all three cases, the research teams capitalised on existing knowledge, which they used to facilitate a partnership debate on issues of participatory engineering, and institutional and procedural design for citizen participation in water resource management and/or regional planning;
– Training and capacity-building in diagnostic/planning processes. In all three cases, the research teams developed and implemented a "training-action" programme designed to build the capacity of both administrative pilots and regional agents from agricultural departments who are responsible for setting up and running multi-stakeholder platforms (see for instance chapter 5);
– Support in carrying out a participatory diagnostic, followed by the formulation of an action plan and an implementation strategy. The researchers help the stakeholders to collectively produce a strategy and an action plan, then to prepare its implementation, taking care to ensure compliance with the principles initially defined (e.g. compliance with the principles set forth by the project, such as transparency, local governance, inclusion & equity, etc.);
– Process monitoring and evaluation and its impact. The researchers assist in setting up a mechanism to "systematically" monitor participation, which is itself partially participatory (see chapter 10), and conduct research on the impact of the consultation process in terms of individual and collective learning and the reconfiguration of power relationships;
– Scientific and technical expertise. In each case, the researchers also carry out (at different stages in the process) complementary studies and expertise on, for example, issues linked to the development challenges identified with local stakeholders (e.g. studies on local industries, on the state of natural resources, agro-ecological experiments, on governance, economic analysis of services, design of information systems, etc.). This expert support is not necessarily planned or anticipated. The researcher accompanying the participatory process may also find him/herself called upon to provide support and expert advice in his/her own areas of expertise.

For each of the projects, the researchers took part in these six stages of the participatory process. However, these stages were not necessarily linear or sequential, and those involved in the process evolved over its course, including the researchers, authors of this chapter. Between experiences in previous projects, training carried out prior to project initiation and the learning acquired as the project progressed, each of us found ourselves in different positions evolving throughout the projects. Table 7.1 provides a summary of all the projects and phases of participation.

Table 7.1. Players involved and researcher participation phases for the three case studies

	New Caledonia	Tunisia	Brazil
Financing	Government	Government, AFD (French Development Agency) and FFEM (French Facility for Global Environment)	Government and AFD (French Development Agency)
Brief objective	Design and implement the country's water policy	Strengthen local governance mechanisms and design and implement territorial development schemes in rural areas	Co-construct a model for multi-level territorial water governance
Regulatory and institutional framework	Development of the country's first water policy, called the "Politique de l'Eau Partagée" (Shared Water Policy)	Democratic transition and policy for the decentralisation of public action (towards the regions and municipalities)	Water policies: work on institutional design to decentralise water management and promote increased cross-sectoral management
Duration	Construction of the action plan and policy guidelines over 12 months (2018-2019), training had already taken place (2016-2018).	Cluster of projects over eight years (2016-2024)	Cluster of projects (from 2018)
Period (stages)	i) Diagnostic (three months) ii) Participatory process (forum + local workshops) (three months) iii) Finalisation of action plans and master plan (three months) iv) Validation by Congress	(i) Methodological development (2016-2017) (ii) Scaling up to six pilot areas (PACTE 2018-2024)	(i) Diagnostic (2018-2019); (ii) Pilot (2020-2023); iii) Loan/scaling up (2024-2029)
Local initiator	Head of Agriculture and Customary Affairs in the Government of New Caledonia	Ministry of Agriculture and Regional Commissions for Agricultural Development	Secretariat for Water Resources and Funceme (Fundação Cearense de Meteorologia e Recursos Hídricos)
Main partners	Operational support by the MISE (Mission interservice de l'eau – Interdepartmental service on water)	French and Tunisian agricultural research and teaching institutions	Secretariats for Agrarian, Environment and Urban Development (sanitation)
Other participants (number)	Customary stakeholders, Municipalities, Farmers, Mining industry, Drinking water manager, State services, NGOs, Consumers group, Citizens (1/600 Caledonians)	Tunisian agricultural administration offices and agencies (4); municipalities (7); civil society organisations (3); citizens (around 4,000 for the diagnostic phase)	Municipal teams, civil society organisations

	New Caledonia	Tunisia	Brazil
Main methodological challenges	Setting-up a methodological approach to ensure the participation of 500 people during a 3-day forum, followed by citizen workshops in the field	Co-designing a methodological approach to reconcile regional planning and large-scale participation	Co-designing a methodological approach to reconcile multi-level water governance and regional planning
Researchers' contribution	Occasional involvement in methodological design and support for workshop facilitation in the five stages	Ongoing involvement in the six stages described above (from set-up to monitoring and evaluation)	Ongoing involvement in the six stages described above (from set-up to monitoring and evaluation)
Participatory methods	Mobilisation and adapting of Cooplage processes/tools	Mobilisation of Cooplage and Co-Obs approaches/tools in a territorial approach (territorial diagnostic, strategic forecasting/vision, planning, implementation, monitoring of territorial dynamics)	Territorial approaches adapted from Cooplage: (territorial diagnostic, strategic forecasting/vision, planning, implementation, monitoring of territorial dynamics)

▸▸ Dynamics and tensions around research stances constructed during the process

In this section, we describe the ambitions and approaches shared by the various researchers involved in the three projects, as well as the different roles that the researchers took on during their projects. The contributions required of the researchers, as described in part 1, are sometimes at odds with the expectations of the participants, whether or not they were made explicit at the outset. As a result, the different roles or positions that the researchers had to assume in these projects often evolved, leading to ongoing tensions and negotiations between the roles defined with the other players (funders, project partners and scientific management), the expected roles as well as each individual's specific position.

Transformative ambitions, at the interface with the political mandate

The researchers involved in these projects all share, albeit with varying degrees or forms of personal commitment, a transformative and democratic ambition. Their aim is to enable the expression of the most diverse points of view, perceptions and interests, and particularly those of the most vulnerable, in order to help improve living conditions and the management of natural resources. To achieve this, the researchers' main ambition and challenge is to ensure a balance of power and to help reduce disparities in the ability to participate in management. In order to help participants (Sen, 1999) and facilitators build their capacity and ensure that what they have learned is sustained, researchers usually offer theoretical and applied training combined with practical activities in the field. Beyond training, it is the "quality" of the participatory process itself, for which the researcher often serves as the guarantor, that allows for a diverse range of voices and interests to come forth, thus guaranteeing a democratic process.

Aside from this shared ambition, each researcher gets involved in the process in a different manner, depending on his/her history in the field and interpersonal relationships with the partners, his/her discipline and research objectives, as well as the meaning she/he gives to the very notion of "involvement".

In the three cases mentioned above, the projects were defined in response to a government request, justified by previous contacts and projects. The researchers helped to develop project aims and organisation, and sometimes negotiated with the funding body (as was the case in Brazil and Tunisia). They positioned themselves at the interface between the transformative political mandate and their own research and innovation mandate. The next step was to design the future course of the project in detail, working with a pilot group to specify what was expected of the various players, the project stages, the resources to be mobilised and how to manage contingencies. In parallel, an analysis of the governance (sometimes included in the subsequent diagnostic) can be conducted to initiate a plan for its adaptation. Here, the researchers provide methodological support and draw attention to specific participatory issues. They also raise their own questions and enumerate constraints (time-frame, publication), as well as establish their legitimacy for the future.

Dynamic positions, with a strong inter-relational dimension

Whatever their own objectives, their original discipline or their skills may be, the participatory co-construction process requires each researcher to adopt an understanding and active approach to the expectations of the participants - project backers, funders, decision-makers and citizens - in order to take stock of the diversity of voices and interests. Participants' expectations evolve over the course of the project's implementation, during the participatory workshops and during process evaluation; although they are different and sometimes contradictory, they are also often concomitant. The researchers' contributions thus evolve over the project's phases, as does the researchers' understanding of the context and the process at work.

Here are a few illustrative examples that demonstrate how partner expectations evolve, intersect and challenge the researcher's stance throughout the transformative process.

– From supporting the participatory process to facilitating it

The researchers support the facilitation of the participatory process by training local facilitators who are acculturated and speak the language, as well as by monitoring and, if necessary, redirecting their activities. To do this, they define a training plan, which is then adapted and fine-tuned with the group of facilitators along the way (see chapter 5). At the request of the participants, the researchers can also act as facilitators themselves for various participatory workshops, ensuring a certain balance in the unequal power dynamics of exchanges between stakeholders; they may also act as workshop leaders to support the process. There is a thin line between facilitation support and facilitation itself, with partners sometimes expecting more in the way of direct facilitation.

Furthermore, as co-pilots of the process, researchers are sometimes expected by local (government) pilots to monitor and deal with any frictions and crises that may arise with stakeholders, or in connection with collateral effects. This involves rapid, contingent analytical expertise on the interplay of stakeholders, requiring direct interpersonal skills (including mediation, negotiation and conflict management) and an understanding of the political and socio-environmental risks. They may therefore find themselves legitimising technical or social innovations that were introduced, or legitimising political decisions that attempt to "correct" asymmetries in participation through these projects.

– Between producing and transcribing knowledge

While, in theory, the researcher simply supports the participatory diagnostic phase, in practice, the pilots often expect them to play an expert role in supporting the production of inventories or comparisons, and studies of past/present dynamics and developments. They may find themselves in charge of study summaries, ensuring their scientific quality and therefore producing original knowledge through their disciplines. In the three projects under consideration, the researchers come from very different disciplines: agronomy, management sciences, geography, sociology and participation sciences. Depending on their discipline, their research objectives and their own publications, they may contribute complementary expertise and disciplinary competencies in addition to their skills in supporting the participatory process. In this way, the researchers themselves produce knowledge that they share with the stakeholders.

At the same time, researchers must also ensure that the diversity of stakeholders is taken into account in the participatory process. They are therefore transcribers of knowledge (Daré and Venot, 2016), i.e. spokespersons for the points of view and representations of the various participants. The stakeholders' points of view may be different, or even in opposition to the researcher's own conclusions; the researcher is thus in a position where she/he must manage possible divergences.

Finally, in the processes studied, the researchers bring their own field experience on what the stakeholders need to mobilise in the process (e.g. stakeholder mapping, systemic modelling, etc.). In this case, the researchers also influence the participatory diagnostic by contributing new methods for collecting, analysing, synthesising and reporting information. This methodological framework influences the participants' approach.

– A dual role in monitoring and evaluating the process

In the three projects, the researchers contributed to monitoring the process and also, in part, to its evaluation. Indeed, the evaluation process makes it possible, on the one hand, to feed into and facilitate the steering or accountability of the process and, on the other hand, to feed into scientific reflection on endogenous evaluation (see chapter 10). This is conducted in part directly, and partly through the use of trained and mentored evaluation managers.

Constructing a stance while in action: a "trial and error" experiment that generates tensions

The researchers facilitating the process may have varying and multiple positions depending on their skills, their personal choices in the face of the explicit or implicit expectations of the funding bodies and project sponsors, the expectations of the participants, the different researchers' own research objectives, the explicit roles within the project, and events. In complex, long-term projects, which often involve the professional mobility of the participating researchers, there is no single project leader. Involvement in any one activity phase may be shared between several researchers, with co-sponsors, co-designers, co-trainers, co-leaders and co-evaluators, with each participating researcher potentially taking on his/her own stance, different from those of his/her colleagues.

In this changing, multi-actor context, researchers can find it difficult to develop and maintain a single stance. Being both "active and reflective", "facilitator and neutral" as well as "expert and referee" generates stress for participants and researchers alike. These tensions are closely linked to the need to interact actively and comprehensively throughout the participatory process, with all the partners—the risks of the process being monopolised by the local steering committees or financial backers is not negligible. In this way, each person's individual stance is built through their actions and involvement as the project progresses, and this construction often remains the fruit of individual experimentation.

From an epistemological and ethical point of view, this typically raises many questions about the relationship between researchers, the steering of the participatory process (decision-makers and politicians) and the participants in that process. During the course of the project, the functional and normative conditions of the research (Checkland and Holwell, 1998) need to be revealed in advance. These contributions and their changes need to be clarified and formalised; they need to be verbalised despite certain risks (Ferrand and Raymond, 2006), so we know "with whom" and why we are collaborating. These changes in stance need to be questioned throughout the process, and researchers need to cultivate a form of reflexivity about their own changes in stance and their contributions.

Illustrative but not exhaustive, the three example projects demonstrate that, even when previously formalised methodological frameworks are in place, this type of project requires expectations to be shared and a certain flexibility on the part of the researchers, who may have to change their stance and their activities during the course of the project. This flexibility, which is necessary for a transformative process, is not without risk for the researcher, who is involved in the long-term process and in the interrelationship with the players and participants.

▸▸ Sharing risks and lessons learned through testimonials

In this final section, we use personal accounts to show how involvement in these processes in support of water governance and policies not only generates tensions for the researchers, who are torn between various expectations, but can at times also place them "at risk", in their interactions with stakeholders and in their role as producers of knowledge. This is not meant to be a comprehensive account. The intention here is to share feedback on experiences and highlight some of what has been learned.

Developing reflexivity while in action

– How can political processes be transformed and analysed?

There are many similarities between carrying out participatory research with the aim of supporting public decision-making and evaluating public policy through a participatory process. The main difference between them is undoubtedly the aim: the evaluation of public policies has a more systematically normative aim, towards supporting decision-making, which is not necessarily the primary objective of participatory research.

Furthermore, the evaluation of public policies, whether participatory or not, is a relatively standardised activity in France. Nevertheless, when participatory research includes citizens and supports the development or implementation of a policy, it is very similar to a participatory evaluation of public policies (Girard and Hassenforder, 2019).

What is very special about the three case studies is precisely the positioning of the researcher, who is involved not only in evaluating the policy in question but also in transforming it and evaluating the process at work. The participatory research process thus generates, on the part of the committed researcher, an evaluation of the public policy that he or she is helping to transform. The need for the researcher to intervene is justified by the initial observation that the policy in question is not working properly. The results, weaknesses or inconsistencies of these public policies can be difficult to explain or make visible by the involved researcher when this leads to criticism of the policies implemented by the government partner itself, which is also the project leader. Furthermore, the process of transforming public policies, even in the sense of democratisation as such, is not neutral. The scientific and technical team may find itself putting forward citizen solutions and demands that run counter to political decisions and expectations, including those from its own project backers/funders.

In the case of Tunisia, for example, sharing the observation made by those behind the PACTE programme, within the Tunisian Ministry of Agriculture, on the excessively limited involvement of local stakeholders in the definition and implementation of natural resource management and rural development policies, the researchers set out to facilitate the development and implementation of a process combining a territorial approach and large-scale participation. By developing methods and tools for diagnostic and integrated planning (i.e. deliberately without constraints on the target sectors), and by strengthening the capacities of regional agricultural services to facilitate the expression of the concerns and needs of local stakeholders, the researchers and their development partners assumed that the programme would be able to generate greater interest and commitment on the part of the inhabitants of the six target regions. This assumption proved to be fully valid initially, with remarkable participation rates recorded during the diagnostic phase (i.e. around 4,300 direct participants in total, 35% of whom were women, and almost 12,000 proposals for action collected from local stakeholders, see Braiki *et al.*, 2022). However, more than a third of the proposals made by local stakeholders focused on sectors that were not eligible for support from the PACTE programme (e.g. transport infrastructure, housing, health, education, off-farm activities, etc.). Thus, many of these actions were then integrated into territorial development plans. Although this result did not come as a surprise to the Ministry of Agriculture, which was heavily involved in the various stages of the process of co-designing and implementing the approach, it still posed a major problem for them. Cross-sectoral mobilisation efforts have been made at central and regional levels, although their success has been limited; local players were made aware, from the outset of the process, of the general conditions for eligibility of the proposed actions, but ultimately, the Ministry of Agriculture now has to deal with plans that go well beyond its remit and, even more so, the technical and financial framework of the PACTE programme. In practice, PACTE contracting authorities have had to deal with major tensions, not only with local players (and sometimes even with regional coordinators) who want to see ineligible actions financed (Hassenforder *et al.*, 2022), but also with funding agencies who are reluctant to modify the financial framework at an advanced stage in the programme.

Over time, these tensions have led to frustration and, to a certain extent, disengagement on the part of some local players, partly invalidating the hypothesis made at the start of the project on the capacity to engage the parties involved.

In the end, such participatory action research approaches, although they involve a highly inclusive co-design process, can sometimes "trap" decision-makers and donors, generate major contradictions between the financial framework and the project "products or outcomes" and, in so doing, highlight major - and sometimes undesirable - imbalances in the power relationships in place.

– How can the participatory process be evaluated and adjusted?

The participatory process can also lead to a modification of the power games at work. The position of certain players is strengthened and legitimised, but this is not necessarily the case for all, or for the most marginal players. The researcher's commitment to the process and his/her desire to transform and democratise make it difficult to objectively evaluate the analysis of power games, or to recognise the failures or limitations of the process in very specific contexts. While the principles of "good" participation are already well theorised, their implementation does not always go according to plan. One of the tensions that the researcher has to manage is precisely the capacity to "give an account" of the limits (or possible failures) of the participatory process and explain the causes, including his/her share of responsibility. This capacity for self-criticism of the process, for which a researcher may be responsible or serve as project leader, is extremely important and may require reorganisation and reformulation of everyone's roles, more training, "course corrections", etc.

For example, the global pandemic hampered the start of the Brazilian project, requiring the work to be launched remotely; initial training could not be properly conducted, thus leading to a participatory process that was inadequately prepared. With a considerable delay, the researcher-project leader, although aware that not all the conditions for analysis and preliminary training of trainers had been met, had to act with urgency, make compromises and launch the process. These compromises in relation to what he had learnt in theory as the "best" way of organising participation had to be made in a highly politicised context with the approach of presidential, legislative and government elections. At the same time, very strong power struggles were emerging between the project's various strategic partners, and the two main Brazilian partners found themselves in a situation of heightened competition. The Brazilian project leader found this start to the co-construction process unsatisfactory. The project leader's ability to provide support was called into question and compromised by this false start. Although the decision to launch the process was not the responsibility of the project manager, but that of his Brazilian counterpart, and although he tried to sound the alarm, to correct the situation and to take a constructive look at what was not working as planned, he did not succeed in satisfying the Brazilian pilot. Paradoxically, however, the project leader's analytical capacity and critical viewpoint were appreciated, as he was asked to take on a new role in the project, as the person responsible for research reports that could feed into the dynamics with the stakeholders and also the interaction with the funder.

This example shows, on the one hand, the compromises that researchers may need to face with regard to the ideal principles of participation, and on the other hand, the limitation of the critical self reflection of a committed researcher. Because of political, financial and time pressures, participatory processes are often launched without all the ideal conditions being met. The attentive researcher, faithful to his/her commitment, can then try to correct the situation, point out the limits, show what is not working, and better advise the pilot - even if the final decision on how to proceed does not lie with him. However, how far can she/he go in criticising without losing his/her credibility, the trust of his/her partner and compromising the whole process, or seeing his/her position in the project threatened? The participatory process calls into question the room for manoeuvre for "committed" researchers or the degree of interventionism that is possible and desirable, and can therefore lead to a change in the researcher's position vis-à-vis the power games that are revealed or emerge during the participatory approach.

The transformative process calls for a rethink of the need for knowledge and the place of expertise

In the three cases studied, the participatory process raises questions about the knowledge needed to make decisions and define an action plan, including the production of knowledge generated by the researchers in charge of the process. Several articles have shown that the need for knowledge can be used, for example, as a lever for negotiation, or as a means of postponing a decision (Bouleau and Deuffic, 2016; Mitroi *et al.*, 2022). Conversely, in the case of the project in New Caledonia, the collective and participatory process sometimes called into question the need for information and the production of knowledge, more specifically on hydrology, river quality and biodiversity. During workshops at the local level, some decision-makers told the researchers: *"I don't need knowledge to make decisions".* Several participants also pointed out the risks inherent in transparency and the transmission of information. If waterholes or springs are inventoried, there is a risk of making them public and, in some cases, making them more difficult to preserve. The case of the preservation of fruit bat nests is fairly emblematic, with a refusal on the part of the customary community to publicise the location of the inventoried nests in order to limit poaching. Understanding the impact of information on the individual and collective behaviours of the participants can thus raise questions about the need for knowledge and its sharing, thus calling into question the fundamental role of researchers as producers of knowledge.

At the other extreme, in Brazil for instance, due to climate factors (i.e. the very high variability of rainfall over time and space being the main management challenge) and historical reasons (i.e. the implementation of a participatory management system for the allocation of water resources in the 157 strategic reservoirs equipped with water level monitoring systems), data is at the core of the allocation system and therefore of management. It would be unthinkable to manage without data. One of the deliverables of the Sertoes project was the design of a Territorial Information System to monitor the 100,000 small and medium-sized reservoirs in the state of Ceara and to incorporate them into a new territorial management model that includes the local level. This objective is in line with the activities of the Brazilian partner, Funceme, which for years has been developing and operating a wide range of expertise to help understand droughts through seasonal climate forecasting, mapping of various environmental factors (soil, vegetation, etc.), as well as assessing the impact of climate variability on water resources and agriculture. It also plays a role in developing decision support systems (DSS) for the water resources sector, and is involved in drawing up emergency plans in the event of drought. The institution's excellent reputation can lead stakeholders at participatory workshops to adopt a stance of waiting for data or a technical or expert solution. Ensuring that all points of view are expressed, without technical knowledge "overpowering" others, is one of the main roles expected of researchers who lead participatory approaches.

Moreover, the researcher's "expert" stance can itself lead to bias in the participatory process, with the researcher bringing his/her own knowledge and expertise to the table. As experts in water and water management, for example, researchers themselves produce knowledge and diagnoses that they share with stakeholders and inject into the participatory process. They have expertise in their own disciplinary field, whether technical or from the social sciences, which may influence their intervention with the participants.

The question of information and expertise in the participatory process can therefore be examined from two angles. On the one hand, it is a question of informing the group, putting it in a position to make an informed decision while retaining the point of view of local knowledge and interests in a participatory process fed by expert knowledge. On the other hand, the aim is to guarantee the legitimacy of the collective decision, even when it does not appear to be the "best" decision, to prevent expert and/or political players from devaluing a solution that does not seem to them to be well argued or scientifically validated. The support approach must therefore help to create the conditions for a rebalancing of knowledge and expertise, including the researcher's own knowledge. This rebalancing is all the more important when the intervention is carried out abroad, by French or researchers from continental France, who may be perceived as representing interests other than local ones, or even as giving lessons.

Lifelong learner: continuous learning through action

All the researchers involved in these three projects share the same observation: their involvement leaves neither the researchers nor the participants "unscathed". The transformative process also transforms the researcher. All three cases required readjusting or adaptations throughout the process, in terms of the way the process was conducted on the one hand, as well as on the skills to be brought in and the needs in terms of research.

For example, in the Brazilian case, project coordination was transferred from the researcher who initially set up the project with partners he had known for a very long time, to a new researcher who arrived in Brazil at the end of the Covid crisis. As the project was delayed and the conditions for an ambitious participatory process were no longer in place, the researcher in charge of the project had to be replaced. A less ambitious trajectory, in terms of participatory actions during the pilot project, was negotiated between the partners, the donor and the various researchers involved in the project, who saw their roles redefined, but also their individual positions evolve in relation to their initial involvement. The adjustments made during the process may have generated frustration and tension for the researchers involved and those who had developed the initial approach.

Participatory projects place the researchers in a rather paradoxical situation of learning as they go, but without always having the opportunity to "sort things out" and do them again or better. This situation can initially lead to self-criticism, with researchers questioning their own ability to lead the process or the value of bringing their own skills into the process. It is the group discussions between colleagues and the sharing of experience, as in this chapter, that ultimately enable these researchers to take a step back, adopt an analytical stance and identify the necessary learning. This learning is necessarily collective, as it is built up with others in a community of practice that enables the multiplication of experiences and meanings given to this type of commitment. The mentoring role that more experienced researchers can take on with regard to younger researchers is also important in learning participatory practices.

Involvement in participatory and support processes is also a privileged learning situation in terms of stakeholder interaction and power relationships. However, although the researcher is able to analyse and see these power plays, he/she also ends up participating (intentionally or not) in these power plays, which evolve over the long term. For example, by helping a government to "democratise" a policy, they are helping to legitimise that government, which may evolve during the process and/or be open to criticism. Over the course of the process, the researcher's commitment may evolve in function of the changes in the balance of power, with some taking a more reflexive stance, being less active or transformative.

Lastly, the process may shape the disciplinary research of certain researchers. The collective process is a place for innovation and creativity of the researchers themselves. Participants often highlight their need for knowledge, expertise and understanding. Conversely, certain needs for expertise or knowledge may be set aside by the participants, raising questions about the positioning and even the need for the skills of some of the researchers involved. Commitment to the process, which is very time-consuming, is often to the detriment of academic recognition, which relies heavily on publications. While this approach does bring us closer to society and decision-makers in the long term, most of the researchers involved in this type of project feel that they need to take some time and step back for more reflective analyses.

⇥ Conclusion

Aware of the scientific and normative stakes of the researcher's involvement in accompanying large-scale participatory processes, in this chapter we have attempted to understand the researcher's stance in relation to the projects, their trajectory in the field, as well as their interactions with other stakeholders (Daré and Venot, 2016; Ferrand *et al.*, 2021). As mentioned by Coutellec (2015), rather than freeing ourselves from these biases of involvement or ignoring them, we sought to make them explicit in order to integrate their scientific and operational consequences. In so doing, we have analysed the tensions and conflicts that arise in the construction and evolution of researchers' stances. The testimonies and feedback shared in this chapter show that, even when we have previously formalised methodological frameworks, involvement in large-scale participatory projects in support of water and regional governance requires a degree of flexibility on the part of researchers, who may have to change their positioning and activities during the course of the project. This flexibility, which is necessary to support transformative processes, is not without risk for the researcher who invests in long-term processes. Local and global conditions, social tensions and the relationships that are (un)forged influence the construction of a scientific stance, not to mention the psycho-social factors specific to each researcher, who also has needs in terms of recognition, integration, legitimacy, security and so on. The research stance is dynamically re-de-constructed, with experience in the field affecting not only the researcher, but also those involved in the field through their questions, formulations and expertise. This type of process is therefore a formidable source of creativity and learning for participants, researchers, funders and decision-makers. These are places for producing and transcribing knowledge, exchanging expertise and local know-how that often bring about simplifications, translations, as suggested by Zwarteveen *et al.* (2021): *"Comparisons across heterogeneous communities sometimes require difficult translations and simplifications. To avoid getting trapped in one single language, we suggest nurturing and thinking with differences, learning from each other's idioms so that no one remains the same as they were at the beginning".*

However, constructing a scientific stance in action and in close interaction with funders, decision-makers and citizens is still often a matter of individual experimentation, a source of tension for the researcher. The accounts given in this chapter, far from being exhaustive, provide a forum for sharing experiences and learning. They show the need to take into account and raise the question of academic recognition of committed research with a transformative aim, particularly in terms of recognition of the specific requirements of such research and the organisation of traceability (see chapter 10). These projects also raise questions about the training of researchers in these approaches, the need for researchers to reflect on their work, and the need to develop tools for clarifying objectives and preparing consultation, as well as monitoring the role of researchers. As mentioned in chapter 9, it is also possible to get the stakeholders to work beforehand, before the start of the participatory process, on who is going to participate at what stages, with what roles, according to what rules and for what outcome. This need for transparency (and clarification of roles) also applies to researchers, especially as they can often be likened to project leaders, and the co-construction dialogue is as important as the final result.

⏩ References

Akrich M., Callon M., Latour M., 2006. *Sociologie de la traduction - Textes fondateurs.* Presses de l'École des Mines.

Alphandéry S., Bobbé P (eds.), 2014. *Chercher, S'engager ?*, Communications, n° 94. https://www.cairn.info/revue-communications-2014-1.htm

Ballestero A., 2019). The anthropology of water. *Annual Review of Anthropology*, 48: 405-421.

Ballon J., Le Dilosquer P.-Y., Thorigny M., 2019. *La recherche en action : quelles postures de recherche ?.* Éditions et Presses universitaires de Reims.

Barnaud C., D'Aquino P., Daré W., Mathevet R., 2016. Dispositifs participatifs et asymétries de pouvoir : expliciter et interroger les positionnements. *Participations*, 16, 137-166. https://doi.org/10.3917/parti.016.0137.

Bouleau G., Deuffic P., 2016. Qu'y a-t-il de politique dans les indicateurs écologiques ? *VertigO*, vol 16 num 2, https://doi.org/10.4000/vertigo.17581.

Braiki H., Hassenforder E., Lestrelin G., Morardet S., Faysse N., Younsi S., *et al.*, 2022. Large-scale participation in policy design: citizen proposals for rural development in Tunisia. *EURO Journal on Decision Processes*, 10, 100020. https://doi.org/10.1016/j.ejdp.2022.100020.

Brun E., Betsch J.-M., Blandin P., Humbert G., Lefeuvre J.-C., Marinval M.-C., 2007. Postures scientifiques et interdisciplinarité dans le champ de l'environnement. *Natures Sciences Sociétés*, 15, 177-185. https://www.cairn.info/revue--2007-2-page-177.htm.

Buono A. F., Savall H., Cappelletti L., 2018. *Intervention research: From conceptualization to publication.* IAP.

Charmillot M., 2021. Définir une posture de recherche, entre constructivisme et positivisme. *In :* F. Piron, É. Arsenault, *Guide décolonisé et pluriversel de formation à la recherche en sciences sociales et humaines.*

Checkland P.B., Holwell S., 1998. Action Research: Its Nature and Validity, *Systemic Practice and Action Research* (11)1. DOI: 10.1023/A:1022908820784

Crémin E., Linton J., Mitroi V., Deroubaix J.-F., Jacquin N., 2018. Quelles alternatives de participation dans les territoires de l'eau, *Revue Participations*, 2/21, pp. 5-36.

Coutellec L., 2015. Pour une philosophie politique des sciences impliquées. Values, finalities, practices. *Ecologie & Politique*, 2/51, pp. 15-25. DOI 10.3917/ecopo.051.0015

Dafflon B., Gilbert G., 2018. *L'économie politique et institutionnelle de la décentralisation en Tunisie: État des lieux, bilan et enjeux.* Éditions AFD.

D'Aquino P., Seck Sidi M., 2001. Et si les approches participatives étaient inadaptées à la gestion décentralisée de territoire ? / And if participative approaches were inadapted to decentralised territorial management? *Géocarrefour*, vol. 76, n°3, 233-239. https://doi.org/10.3406/geoca.2001.2561

Daré W., Venot J.P., 2016. Dynamique des postures de chercheurs-engagés: retours sur la participation dans les politiques de l'eau au Burkina Faso. *Anthropologie et Développement*, (44), p. 149-178.

David A., Hatchuel A., Laufer R. (Dir.), 2012. *Les nouvelles fondations des sciences de gestion : éléments d'épistémologie de la recherche en management.* Presses des Mines, pp.278, 2012, 978-2-911256-90-5. https://hal-mines-paristech.archives-ouvertes.fr/hal-00748217

Ferrand N., Hassenforder E. Aquae-Gaudi W., 2021. Quand les acteurs modélisent ensemble leur situation, principes ou plans pour décider et changer durablement, en autonomie. *Sciences Eaux & Territoires*, 35, 14-23. https://doi.org/10.3917/set.035.0014

Ferrand N., Raymond R., 2006. De l'observation à l'intervention dans les processus de gestion ou Faut-il verbaliser les sens interdits ? *In : Traces, Enigmes, Problèmes : Emergence et construction du sens* - 13èmes journées de Rochebrune : Rencontres interdisciplinaires sur les systèmes complexes naturels et artificiels. ENST 2006-S001. pp. 163-175.

Girard S., Hassenforder. E., 2019. *EvalPart2 : deux expériences d'évaluation citoyenne de la participation dans la Drôme : SAGE et PLU.* 13èmes Journées Françaises de l'Évaluation «Évaluation et démocratie : les nouveaux territoires de l'action publique», Jun 2019, Bordeaux, France. https://hal.inrae.fr/hal-02609640

Hassenforder E., Braiki H., Lestrelin G., Faysse N., 2022. Transaction sociale et participation : Un projet de recherche-action sur le développement territorial agricole en Tunisie. *Pensée plurielle*, 55, 175-190. https://doi.org/10.3917/pp.055.0175

Herreros G., 2009. *Pour une sociologie d'intervention.* Érès. https://doi.org/10.3917/eres.herre.2009.01

Mazzocchetti J., 2007. Retour réflexif sur deux expériences contrastées de recherche impliquée. *Politiques sociales*, 3 and 4, 25-41.

Mitroi V., Deroubaix J.-F., Tall Y., Ahi Kouaido C., Humbert J.-F., 2022. Rendre compte de la dégradation des milieux aquatiques. Le rôle des savoirs dans la mise en place des politiques de protection des ressources en eau en Afrique subsaharienne, *Géocarrefour* [On line], 96/1. https://www.doi.org/10.4000/geocarrefour.19353

Olivier de Sardan J., 2021. *La revanche des contextes : Des mésaventures en ingénierie sociale en Afrique et au-delà.* Karthala.

Sen A., 1999. *Developpement as Freedom*, Oxford: Oxford University Press.

Weber M., 1965. Essai sur le sens de la "neutralité axiologique" dans les sciences sociologiques et économiques, *In : Essais sur la théorie de la science*, Paris, Plon, p. 475-526.

Zwarteveen M., Kuper M., Olmos-Herrera C., Dajani M., Kemerink-Seyoum J., Frances C., *et al.*, 2021. Transformations to groundwater sustainability: from individuals and pumps to communities and aquifers. *Curr. Opin. Environ. Sustain.* 49 (2021), pp. 88-97, 10.1016/j.cosust.2021.03.004

Supporting participatory policy making with an integrated digital platform: The e-CoOPILOT approach

Nils Ferrand and Samuel Tronçon

As described elsewhere in this book, participatory policy making in multi-stakeholders intersectoral and multi-level contexts is an intricate process that requires addressing various needs, constraints and steps. Beyond in-presence protocols (i.e. processes in which participants are physically present), digital solutions provide opportunities for social extension ("massification"), autonomisation and compliance with current practices. But some participants can be reluctant to engage in a long, complex and sometimes doubtful pathway. It therefore appears crucial to support them step after step in the decision procedure, to value and share all knowledge produced on the way and thereby to improve the efficiency of participation. This is what the e-CoOPILOT approach presented in this chapter aims at doing, using artificial intelligence (AI) solutions.

In terms of public policy support and transitions, the ultimate goal of the approach is to value the large experience of the CoOPLAGE tools and case studies, to transfer it in a generic platform open for all stakeholders, thereby giving them the capacity to design, pilot, participate and evaluate some integrated participatory processes. The e-CoOPILOT approach aims at proposing solutions beyond the existing large set of participation platforms with a focus on the global CoOPLAGE decision cycle (see chapter 2). The approach values the transversal role of participatory modelling (see chapter 13), supports the process of "participatory engineering of participation" (see chapter 9) and implements its following steps. As such, it is intended as a coherent "companion" to participatory processes managers and participants, which should strengthen the actual use of participation in democratic decision-making, and foster trust between citizens and institutions. These goals raise several research and design questions, mainly related to the integration of steps and tools in the procedure, and to the capitalisation of knowledge.

We discuss in a first part the target implementation context and a reference use scenario which shaped the design of the e-CoOPILOT platform. In the second part of this chapter, we describe the structural choices and the architecture of the platform. Finally, we discuss ongoing evolutions. This chapter is structured as a design document and not as a scientific contribution, which will come later in the experimental phases.

⇥ Coping with procedural needs and constraints in the engineering of participation support

Contemporary social and environmental challenges require severe transitions. Such transitions cannot be achieved without a coherent and protracted commitment of all stakeholders. Engineering participatory processes, as a global decision cycle including all actors, is a complex task requiring multiple skills, integration of many decisions' steps and tools (see chapter 2), and a protracted management of participants, tasks and products. In this chapter, we define the "engineering of participation" as the design and management of a participatory process, including the protocol, methods, tools and regulations used in its implementation. We specifically consider digital solutions which can be designed and extended to support the engineering of participation. Two questions structure this design:

– What are the essential steps and needs of stakeholders and institutions in participatory processes for socio-ecological sustainability?

– What are the added requirements for supporting participatory processes by digital means (online, mobile)? And how do these digital means combine with physical participation?

Various approaches to participatory processing of transitions

Various approaches exist in the literature to support transition steps and needs (Koning *et al.*, 2021; Hyysalo *et al.*, 2019; Fet and Keitsch, 2023; Loorbach and Rotmans, 2006; Loorbach, 2010; van de Kerkhof and Wieczorek, 2005). They can help us answering our first design question above. In the Transition Support System developed in Wageningen (Dijkshoorn-Dekker, 2018), five steps are considered: urgency, scenario analysis, in-depth analysis, insight into future directions and impact evaluation. These steps can be repeated. The Transition Support System mainly focuses on prospective methodologies, and emphasises sequences of visioning and backcasting. In Halbe and Pahl-Wostl (2019), the four steps are: problem and actor analysis; participatory modelling with causal loop diagrams; analysis of learning objects, subjects, contexts and factors; and integrated governance system analysis. In the latter, the role of participatory modelling and system analysis is stronger.

Based on several case studies introduced in this book, and driven by water management issues, we have established a different analysis and protocol in the CoOPLAGE framework (see chapter 2). Its decision loop aims at being comprehensive, and transcribed in terms of transition steps and needs, includes the steps in table 8.1.

This generic approach can be, in principle, structured and managed without any participation, by combined intervention of experts and decision makers, followed by a transfer and adoption phase toward other stakeholders. In computer-based approaches, this is the usual approach of decision support systems, where the deliberative side is reduced. However, in the CoOPLAGE posture we aim at including all stakeholders in the various phases of the decision-making process.

This brings us to the second design question: how to adapt these steps when enforcing participation, especially using digital means in combination with the classical, presential, social process?

Table 8.1. Decision steps and their participatory version

Decision step	Description	Participatory implementation
Preparation	establishing the conditions, plans and rules of this participatory process	Participatory design of the decision (hence participation) procedure & organisation
Information & training	Providing the required information & capacity to participants	Mutual information, focus groups, mutual training by specialised groups
Diagnostic, baseline analysis	building and sharing situation analysis for all dimensions (environmental or social) and all scales	Participatory observation & diagnostic, participatory modelling
Prospective thinking, visioning	Designing and using scenarii for the future	Opening prospective thinking to all stakeholders & citizens
Preferences, goals, criteria	Setting the transition or transformative goals or principles	Participatory deliberation on goals and criteria
Monitoring and evaluation + piloting	Setting a monitoring, evaluation and adaptive management plan for this transition process (based on the transformative goals)	Participatory design of monitoring and evaluation – co-steering of the process with a participatory steering committee – co-management of the process
Actions & planning	Building alternative transition action plans composed of sub-actions, coherent and efficient	Open deliberation on actions and plans – participatory planning
Prioritisation	Selecting one action plan and committing stakeholders to it	Participatory selection or vote, large scale formal social commitment to the process
Implementation	Designing an implementation plan and operationalising it	Shared implementation – extended contribution & work

The general answer is simply to extend the group of stakeholders enacting the steps, which leads to specific adaptations which are introduced in the third column of table 8.1. These steps have to be made coherent and incremental.

A reference use scenario

In order to share a comprehensive design and adapt it with the policy holders on one side, and with the platform designers on the other, we have built an abstract reference scenario, in the form of a target use case (box 8.1). It was originally established for the European project SPARE (Ferrand *et al.*, 2017) for river ecosystem services management, and has been adapted to transition processes.

This scenario provides an overview of the target use, based on process' implementation actually occurring in non-digital processes. The question is how such process can be supported by a digital platform, and with which hybridisation between the digital and the physical implementation.

Box 8.1. Reference scenario for a target use case

In this challenging period, a local authority (LA) decides to start a transition planning process (TrP). A process manager (PM) is designated. She identifies and gathers a small pilot group (PG) of eight persons, made of diverse representatives who can help her animating the process. Including an expert, the PG recommends that an ex-ante evaluation is made on a population sample to ensure future comparative evaluation.

Using social media, they communicate widely to the population to inform them about the launch. Volunteers are already invited to register for future works, meetings, etc. Through a dedicated web and mobile application, everyone is invited to propose participatory actions (Participatory action proposals, PAP) to the LA: how they can contribute to the decision, i.e. how citizens should be associated, which rules.

A citizen assembly is gathered with the volunteers. They can access methodological training using an online training course (MOOC). After exchanging with experts in participation, they discuss the PAP and decide the plan and rules for participation. They also decide how the process can be monitored and should be evaluated. The LA and the local politicians also contribute and provide their vision. The draft participation proposal is made fully public and comments are welcome. A final participation program (PrePar plan) and a charter are published and signed by the main representatives. The process can start.

Through a mailbox delivery and by internet, every household (and tourists in their residence) receives an Observation and Knowledge Kit (OKK): a set of simple and robust cards with transition awareness, a socio-environmental monitoring form (with participatory mapping), and preference survey. An open mobile application includes the same. An OKK challenge is organised, with symbolic awards. People (and schools) can travel the surroundings, collect questions and data and share them through the LA. All these data are used by LA with experts to produce a participatory diagnostic, including situation and revealed preferences. OKK public sessions are gathered where people can meet to discuss their observations. Distributive justice dialogue is also facilitated (with JustAGrid protocol): people can express what they consider to be fair in terms of resource and effort sharing.

Smaller groups (and classes again) are invited, with a facilitator, to build models of the territory, and the possible transition pathways, including ecosystems, economy, exchanges. Using an adapted version of the Wat-A-Game toolkit, they obtain all together a general local model where different options and scenarios can be tested through role-playing game sessions (participatory simulation). The model calibration is improved by experts. Several copies are produced and distributed to stakeholders for future uses.

With this model, groups can reassess and challenge their OKK diagnostic.

LA aggregates all results and proposes a draft synthesis which is published and shared. After feedback, a final version of the diagnostic is officialised.

All sessions and stages have been monitored and evaluated with an autonomous contribution of some participants.

In this phase, every household receives a new kit: the Option Proposal for Transition (OPT) part of the CoOPLAN set. Everyone can propose action for, around, and about the transition in the territory. They can send it on paper or share by internet or an application. For each OPT, they have to think about who, what, how and why. All OPTs are published on the LA site, under categories. A Market Place phase is started where people can meet to comment and improve the OPTs.

With this set of updated OPTs, volunteer groups are invited to weave complementary OPTs in Transition Integrated Management Strategies (TIMS) using the CoOPLAN methods. They assess feasibility and efficiency, and a dialogue with experts is organised to criticise and assess the TIMS. All TIMS are published and shared and comments are invited from everyone. A large public dialogue is organised to make summary of all comments.

All sessions and stages have been monitored and evaluated.

Everyone is invited to an official distributive vote about the five final TIMS (physical and electronic vote) with an allocative judgment. They have five points to allocate to all TIMS. At the end of this process, the winning TIMS is designated and made public. It will be implemented.

In a last stage, a steering group will manage and adapt the implementation.

Assessing needs and constraints for an integrated participation platform

The previous scenario is quite comprehensive in terms of features and steps. Such users' needs can be tackled by two ways: through individual tools responding to each step, or through an integrated platform. The latter has the advantage of unifying access, data management and incrementation of contents. Such participatory platform has to cope with the usual requirements of accessibility, user friendliness, transparency and robustness. But other challenges appear when dealing with the global decision cycle in a multi-stakeholder's context:
– addressing the diversity of needs of the various users' categories: process managers, facilitators, analysts, participants;
– unifying data production and knowledge management throughout the platform;
– managing asynchronous use;
– assisting users in their track of use;
– maintaining the same transparency as for the material version while facilitating the process;
– aligning with other classical social networking practices.

Meanwhile, and as addressed in the next part, the digital access and process is not separated from the physical encounters of participants. Various conditions can be encountered among fully digital asynchronous, synchronous parallel, synchronous with users' groups gathered either with one or many digital access, fully physical session supported by one facilitator and one platform access.

▸▸ Milestones and results toward an integrative computer assisted participation

E-CoOPILOT was designed as an online service which aims at supporting participatory process managers, policy makers and all participants in co-designing, piloting, using and evaluating their participatory decision processes, for their various needs and stages attached to socio-ecological transitions, in an integrated and coherent manner.

In this part we assess some existing platforms and consider their response to the same issues, while addressing the global decision loop.

Features and models: similarities and breakups

We wonder here to what extent e-CoOPILOT constitutes a proposal that is really different from existing platforms supporting participatory processes, and what this specificity can contribute to the quality of the processes involved. Considering the existing platforms and tools as referenced in the compendiums[1], the features generally used for comparison include[2]: "assessment of ideas, collaborative budgeting, collaborative drafting, commenting, conversation, debates, events, forums, guided tours, idea submission, mapping, meetings, messaging, moderation, network/graph mapping, notifications, petition, polls, preference and prioritisation, project timeline tracking, proposal splitting and merging, question authorities, recommendation engine, register volunteers, sentiment analysis, sign-up forms, SMS tools, surveys, translation, transparent survey results, verified participation, photo and video management, voting". But a wise and complete comparison must also include the underlined models of decision process and actors, and not only the main functions the platform is offering, considering the model as an operational definition of what a method can give the ability to obtain as empirical and/or social results.

Facing the quantity and the variety of existing platforms, we decided to compare the e-CoOPILOT design with Decidim (Barandiaran *et al.*, 2018), Assembl[3] and CitizenLab[4], selected for the diversity of their features and their ever-increasing number of users and applications (Table 8.2).

Decidim and Assembl were widely inspired and initially funded under the European program CAPS[5], which triggered their emergence on the basis of communities' needs. As such the initial main features were: structuring and supporting the participants' groups, structuring the problem space, supporting debate and deliberation, voting and dissemination. CitizenLab seems to have a very similar underlying model, even if guided differently. The platform was made for a first participatory process and progressively improved as experiments progressed, all along in accordance with the UN goals for sustainable development.

E-CoOPILOT is based on an aggregated analysis over the digital participation tools (RMCPART, 2020), with a slightly more detailed classification: administration of the participatory process, structuring and organising participation, sharing documents and supporting debate, diagnostic and data collection, collecting citizens' proposals, choosing and voting, funding an action.

Regarding the conceptual models, the actor's and integration models (how elements are related) of these platforms are not obvious. Their capacity to cope with the complexity of social processes and the induced action plans, i.e. interlinking issues and proposals with a situation model, although discussed (Barandarian, 2018), is not established

---

1. We used two compendiums: PeoplePowered platform, 2023 (49 platforms referenced) https://fr.people-powered.org/2023-planning – Digital participation platforms research. Airtable. Accessed april 29, 2023, https://airtable.com/shrxxpcHHnMc1xZSx/tblELFP9tGX07UZDo – and Participedia, 2023 (28 methods with software) https://participedia.net/.
2. Categories and terminology are extracted and adapted from the classification of PeoplePowered analytics, ibid.
3. Accessed may 2, 2023, https://www.bluenove.com/en/offers/assembl/
4. Community engagement platform. Accessed may 2, 2023. https://www.citizenlab.co
5. Community Awareness Platform for Sustainability (CHIC, 2018).

Table 8.2. Comparison between platforms

Platform	Origin and model	Common features	Specific features
Assembl	Based on community needs CAPS model	Users and groups admin Actions proposal	
Decidim	Based on community needs CAPS model	Debates Deliberation Dissemination	Budgeting Events surveys Mapping
CitizenLab	Bottom-up from projects needs UN sustainable development goals		Budgeting Events surveys Mapping Analytics
e-CoOPILOT	Based on an empirically proven workflow CoOPLAGE protocol (this book)		Resources/impacts evaluation Events Surveys Mapping Analytics Diagnostic Process design Process management Roles management Endo-evaluation Online course

CAPS: Community Awareness Platform for Sustainability (CHIC, 2018).

in the available material. It seems to be left to the deliberation of participants. The main and classical processing of interdependencies is through the use of semantic classification and machine learning applied to the flows of participant's contributions. The minimal matching model is based on similarity index (of interest, position), but not reconnected to a normative model of a target socio-environmental situation, as required for supporting the transition of territories.

In the CoOPLAGE approach, introduced in this book and grounding the e-CoOP-ILOT design, the integral approach of participatory decision-making is supported by a global theoretical model consisting in four parts:
– a procedural model of a recommended decision cycle,
– an actor, action and plan model,
– a situation model,
– a co-constructed normative model for the endo-evaluation.

Consequently, the very distinctive feature of e-CoOPILOT, in comparison with the other digital participation platforms, is not only to be structured under these four conceptual models, but above all it is their co-integration which produce a strong iterative and progressive workflow, coupling the tools for all the decision steps needed, and consequently the focus on the design and steering of the participatory process itself.

As an intermediary conclusion, we assert that e-CoOPILOT should overcome the current limitations of the other platforms which often gather separated action support for debate, budget, propositions, without a real procedural engineering nor an under-lined model of social transformation and decision. And in the wake of the CoOPLAGE approach applied to a digital support, we consider e-CoOPILOT will foster the coherence of both the process and stakeholders' engagement, and develop trust between parties by the specific relationship it establishes between managers and participants in the instantiation, monitoring and evaluation phases of the processes.

From physical constraints to digital opportunities

Considering the e-CoOPILOT design as a transition from physical methods to digital implementation, we can wonder about this transition process itself, and the impacts on the theoretical model: its implementation, its adaptation, its revision.

The platform was initially designed as a proof-of-concept for a digital implementation of the CoOPLAGE physical tools. In this respect, the main question that arises is twofold:
– Is the implementation consistent with the model? This amounts to ask whether the conceptual specifications are respected?
– Does the implementation make it possible to complete or revise the initial model, and therefore, to modify in return the CoOPLAGE method in its physical implementation?

The implementation was carried out in several iterative steps. Based on an initial data structure matching the CoOPLAGE UML model, we designed the processes and inter-actions necessary to feed, query and navigate in the model. Then, we implemented the "graphical" tools for collaborative work which notably allow a synoptic view of the work carried out by the participants (diagrams, drawings, plans, etc.).

In a second stage, we added some components made necessary for the actual implementation. They fall into two categories. Some are purely technical, relating less to the conceptual model than to the relational model. More interestingly, others are quasi-conceptual elements absent from the initial logical model, which question whether they are real objects, i.e also relevant in the physical "version" of CoOPLAGE, or if they only belong to the digital "version" of the method (figures 8.1 and 8.2). As such, this implementation of CoOPLAGE raises many particularly interesting questions. For example, the representation of the process has evolved throughout the modelling. The users' status in the computer version enforces:
– complete traceability of the users' actions in the model, which is impossible to achieve exhaustively in an empirical process, whereas it is "by design" in the digital model;
– the emergence of a "digital twin" of the participant, specific to digital use, which will be defined both in relation to its categorical definition (age, gender, territory, social categories…) but also by the actions it carries out (proposals, debates, various inter-actions…), and can be used later to support a personal assistant agent (Ferrand, 1997).

Using the digital platform may have consequences on the initial actor and decision model of CoOPLAGE, with the following issues:
– Is the global model respected in the digital version or is it modified?
– Is the adapted model from the digital version responding to extra needs or constraints?

The physical version of CoOPLAGE was implemented face-to-face with groups of variable sizes (up to 5,000 persons), in coordination or not with other groups following the

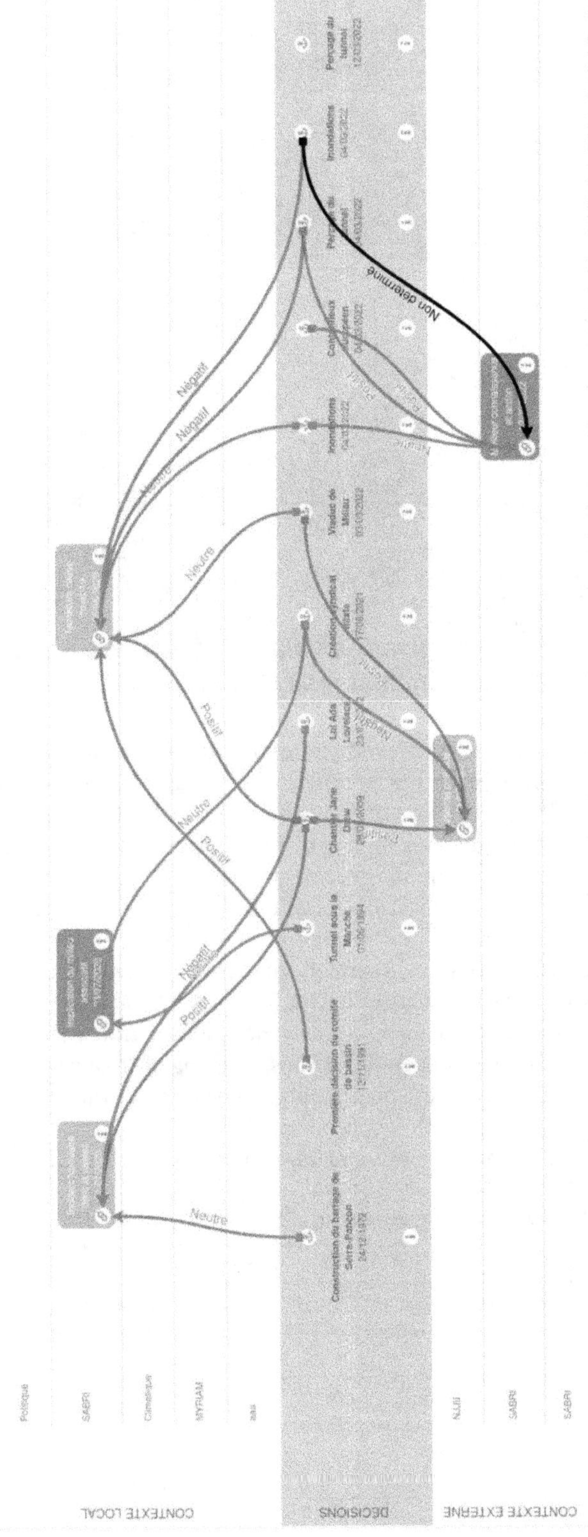

Figure 8.1. A view from the diagnostic module in e-CoOPILOT

A view from the diagnostic module in which we can relate decisions to their corresponding causalities or impacts. All of these objects and relations constitute a graph of the analysed situation.

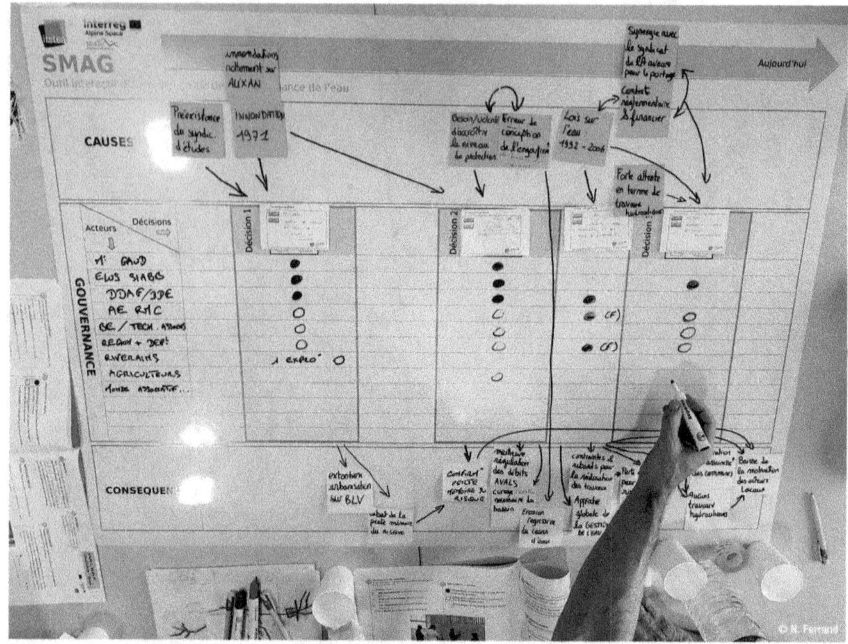

Figure 8.2. The diagnostic phase in the physical CoOPLAGE process

same protocol in different territories and at different times. The main difference with a process carried out on Internet lies in the fact that users can collaborate asynchronously and remotely (Table 8.3). The first version of e-CoOPILOT was designed to be used as a digital medium in the same type of context: a face-to-face group in one place, a facilitator who projects the main interface on a screen (for example a visualisation of the actions proposed by participants) and users on individual computers that perform actions in the interfaces. Quickly, new needs appeared with the possibility of carrying out the same type of session but only remotely, therefore with people connected to the platform simultaneously, which implies having real-time notifications of the actions of the participants. Finally, a third use, asynchronous and remote, which reveals new needs both in terms of interactions (notifications, tracking of changes, alerts), but also in terms of interfaces, because users must be able to progress at their own pace, and it is therefore necessary to allow both synchronisations (collective times for deliberation for example) and personalised routes (for users wishing to go further or more in depth).

Table 8.3. Four types of uses of a digital platform in a participatory process

	Synchronous	Asynchronous
Face-to-face	A projected interface, participants on their terminals to a single work interface.	Separate groups, which contribute at different times to different stages of the process.
Distant	A control interface for facilitators, users connected to the same interface at the same process step	A management and monitoring interface for facilitators, a monitoring interface for users, everyone progresses at their own pace

The transition to digital therefore makes it possible to vary the modes of use, and therefore also the modes of animation of the process, which has influences on the initial model of CoOPLAGE and raises many questions on the way of designing and animating the processes.

Furthermore, it questions the nature of the decision process itself, and the endorsed actor's model. In the physical version, participants meet and influence each other directly, have a dialogue and the deliberation is the base of the decision process, albeit a structured methodology is used. In this digital and asynchronous use, the tool and its procedure may prime on the deliberation which is only a second level, a supportive layer. Meanwhile "others" are disembodied, and the coalition or coordination processes are mainly based on the positions taken with an "in-person" attribution much less obvious.

An operational paradox was that while e-CoOPILOT was being developed, a very large scale face-to-face process had been organised in Tunisia (see chapter 5) which also led to its own parallel procedural adaptations. This obviously has a significant influence on the type of interfaces to be developed according to the different methods, the complexity of development, and in particular about the proceduralisation:
– the possibility or not to advance in the workflow step by step, by closing the steps already completed which remain only consultable, with an influence of the actual steering of the relational procedures,
– the management of various user profiles, depending on their motivation, the intensity and extension of participation, their digital skills and the accessibility of the tools (excess of participatory requests can mislead some users by complicating interfaces).

These questions, initiated by the design of the e-CoOPILOT tool, constitute also operational challenges that need to be addressed, and also true research questions at the crossroads between the science of participation and computer science. In particular, they may require the development of automated assistants to support participants and follow various protocols, adapted to the participation goals and the participants' expectations.

Extended features and constraints to be considered

To strengthen the specification, and based on policy support applications in many countries (detailed in other chapters of this book), we address hereafter some additional insights on some expected functions or targets. They result from several design workshops.

First, regarding the use conditions in real policy settings, a key issue is the level of adoption and integration of the digital tool by all stakeholders in their process. The related requirements include a dedicated pre-training, with easy online access, and a sensitisation to participatory decision-making. The underlined decision model should be aligned with their actual decision cycle, or should be fully customisable. Complementarity with the physical participatory process is crucial, by mirroring the methodologies, and supporting also the physical sessions (their organisation, guidelines, material and monitoring), because even if we can imagine full online participatory processes, the digitalisation does not, and certainly must not, aim to eliminate the physical modalities of debate and cooperation between actors, whose issues and practices are nevertheless anchored in the physical world.

The same stands with the legal context to comply with. Accessibility and low cost are required for massification. Meanwhile, adoption could be fostered by raising awareness of other use cases by various groups, and building a community of practice among stakeholders.

A convincing proof of concept would be based on tests and validations in real and large policy contexts. Process managers should be supported continuously; and robustness and on-time-responsiveness are crucial when we talk about processes bringing together hundreds if not thousands of potential participants. From a scientific and operational perspective, systematic comparisons between equipped and no-tool control situation are required, but as for other experiments, the relevant conditions are extremely difficult to organise.

Secondly, for the content of the platform, the main constraint is on the coupling and integration of the tools. Information has to be cumulative and incremental, and the conceptual model for elements (e.g. actions, actors, contexts models in CoOPLAN and WAG) to be unified and shown. Participatory modelling being the core principle of CoOPLAGE it must structure the users' pathway and the coupling. A transparent monitoring and evaluation process can provide such continuous insight, and clarify the steps' rationales.

Debating is often considered as the core of participation, although here the decision model primes. The attached online debate facility must allow to render the assertions in the other tools, or use any tools' components in the debate.

To feed and assist the process, the coupling with external data or knowledge sources, as well as with AI devices, has to be engineered: searching for relevant data or sources by content analysis, or providing a contextualised assistance based on the collective process or the deliberation model using speech acts sequences, thematic tracks, coalition analysis (as in the Communities of Assistant Agents of Ferrand, 1997).

To be efficient, it should help triggering and sustaining participants' commitments by: transforming contributions into formal intentions, sharing and archiving interactions, and supporting long-term social commitments with the help of social networks.

Finally, considering the main target device, access and use of mobile solutions has to be granted, with specific adaptation for mobile use (especially for the large tables required in the integration phases), with contextualised and localised approach, and with citizen-to-citizen local matching and dialogue.

▸▸ Conclusions

While the CoOPLAGE tools have been extensively used and validated in field applications, e-CoOPILOT is still in a beta stage and should be tested in practice. Most of the issues and needs addressed in this chapter are still to be developed and/or optimised. The essence of this chapter was to elicit the potential use conditions, the position in regards to other platforms and to draw the current development pathway.

E-CoOPILOT is however already a comprehensive platform including non-classical functions for digital participation, like procedural design and steering, explicit models for actors and plans, enforced role of monitoring and evaluation. The CoOPLAGE community expects e-CoOPILOT to become a major instrument for large scale social dissemination of the good practices and impacts of the material tools. Therefore, the user requirements and the interface will undergo an improvement process and a specific adaptation to more generic transition protocols.

Further development is focused on direct support to participants, including mobile use, and integration of data science and AI techniques using the principles of assistant agents' communities (Ferrand, 1997).

As we have argued, assisting users in multi-level policy processes is the key purpose of this design. However, there is a duality between features (what we can do) and models (what we can expect from our actions). Assisting users is a new capability, which may require reviewing the whole conceptual process initially inspired by the physical version of CoOPLAGE.

Obviously, the first capability is to capture, protect and value data that make it possible to represent practices, expectations and assessments from the participants, as in most contemporary user platforms. This goal is focal in any participation: an equal attention paid to everyone's contributions. Once this primary need has been fulfilled, the second need arising is that of interaction between participants, through discussions, deliberations, comments, cross-evaluations. We consider that there can be no real collective construction without debating, and without debating all the stages of the participatory process itself. A third capability refers to the representation of the information and its accessibility. This objective responds to the obvious need for transparent participation, open to all, explainable and assessable. Finally, the fourth objective is to make the process itself manageable, through its stages, its organisation and up to its own end. However, this manageability is only possible if the protocol is formalised, comprehensive, following identified stages, with a clear finality, of which all the participants can be aware and committed to. In particular when they agree to delegate this steering to a restricted group, or to the system. Trust in the system is critical therefore.

A fifth objective is, however, to support the participants in their participatory journey by presenting them the different possibilities available to them at each stage, with pedagogy and using all the means at their disposal to clarify the options and decision in a completely transparent and reliable manner. It may seem to stem from digitalisation, a solution against the isolation of digital actors. It is in fact an integral part of the very process of citizen participation, driven by ethical principles. This function, generally fulfilled by humans who support the physical processes, is essential, and cannot be absent from a digital device in which users can operate independently. In other words, there cannot be digital autonomy without offering the possibility of guidance in the process, its stages, and in the complexity of the information provided: non-binding guidance which must be subject to the choices of the participants, and which must guarantee the possibility of an autonomous exercise of participation excluding takeover by a third party, obfuscation or influence. For these last conditions, we know well that digital technology offers opportunities, thanks in particular to the control it allows over data, while opening up new potential risks which must be investigated and managed.

As we can see, digitalisation and the conceptual unfolding of the issues to which it must respond are only the magnifying mirror of physical participatory issues, especially when we take as a basis a common underlying model. Bounded to a list of functionalities or a focus on opinions' collection, digital participation technology may be disappointing at best and dangerous at worst. The importance of proceduralisation, of a complete and integrated global model, and of supporting participants is not only an operational condition of these systems, it is also and above all an ethical condition which ensures transparency in the core function of the participatory process.

⇥ References

Barandiaran X., Calleja-López A. Monterde A., 2018. *Decidim: political and technopolitical networks for participatory democracy*. Decidim's project white paper, pp. 7-26. Retrieved may 2, 2023. http://ajbcn-meta-decidim.s3.amazonaws.com/uploads/decidim/attachment/file/2005/White_Paper.pdf

CHIC (2018). *D.3.5: CAPS Good Practices Report – Final version*. Passani A., De Rosa S. (coord.) Retrieved may 2, 2023. https://ec.europa.eu/research/participants/documents/downloadPublic?documentIds=080166e5be3b6b2e&appId=PPGMS

Dijkshoorn-Dekker M., 2018. *Transition Support System Approach*; Wageningen Economic Research: The Hague, The Netherlands

Ferrand N., 1997. *Modèles Multi-Agents pour l'Aide à la Décision et la Négociation en Aménagement du Territoire*. Thèse de doctorat en informatique de l'Université Joseph Fourier. Grenoble : UJF. https://theses.hal.science/tel-00003562

Ferrand N., Noury B., Girard S., Hassenforder E., 2017. *Initial Guidelines on Stakeholders' Engagement and Year 1 Participatory Process in the PCS*. SPARE WPT1 D.T.1.1.2 Final report. Accessed may 2, 2023. https://spare.boku.ac.at/index.php/en/about-spare/initial-guidelines-on-stakeholders-engagement-and-year-1-participatory-process-in-the-pcs

Fet A. M., Keitsch M., 2023. Transition to sustainability. *In*: A. M. Fet (Éd.), *Business Transitions: A Path to Sustainability: The CapSEM Model* (p. 215-222). Springer International Publishing. https://doi.org/10.1007/978-3-031-22245-0_21

Halbe J., Pahl-Wostl C., 2019. A methodological framework to initiate and design transition governance processes. *Sustainability*, *11*(3), 844. https://doi.org/10.3390/su11030844

Hyysalo S., Marttila T., Perikangas S., Auvinen K., 2019. Codesign for transitions governance: A mid-range pathway creation toolset for accelerating sociotechnical change. *Design Studies*, *63*, 181-203. https://doi.org/10.1016/j.destud.2019.05.002

Koning S. de, Haas W. de, Roo N. de, Kraan M., Dijkshoorn-Dekker M., 2021. *Tools for transitions: An inventory of approaches, methods and tools for stakeholder engagement in developing transition pathways to sustainable food systems*. Wageningen Marine Research. https://library.wur.nl/WebQuery/wurpubs/577270

Loorbach D., 2010. Transition management for sustainable development: A prescriptive, complexity-based governance framework. *Governance*, *23*(1), 161-183. https://doi.org/10.1111/j.1468-0491.2009.01471.x

Loorbach D., Rotmans J., 2006. Managing transitions for sustainable development. *In*: X. Olsthoorn, A. J. Wieczorek (Éds.), *Understanding Industrial Transformation: Views from Different Disciplines* (p. 187-206). Springer Netherlands. https://doi.org/10.1007/1-4020-4418-6_10

RMCPART, 2020. *Quelle stratégie participative pour la gestion locale de l'eau avec les citoyen? 4. Participation et numérique*. Aucante M., Hassenforder E., Ferrand N., (eds.). Rapport d'étude pour l'Agence de l'Eau Rhône Méditerranée Corse. retrieved may 2, 2023. https://www.eaurmc.fr/upload/docs/application/pdf/2020-09/fiches_participation_numerique_-_irstea_aermc_-_10-06-20.pdf

van de Kerkhof M., Wieczorek A., 2005. Learning and stakeholder participation in transition processes towards sustainability: Methodological considerations. *Technological Forecasting and Social Change*, *72*(6), 733-747. https://doi.org/10.1016/j.techfore.2004.10.002

Part 2

Engineering and Evaluating Participatory Decision Processes

Chapter 9

Engineering participation: Preparing and designing a participatory process

Nils Ferrand, Emeline Hassenforder and Sabine Girard

In concrete terms, participation engineering involves thinking about the objectives, design, choice of methods, implementation, and monitoring and evaluation of a participatory process. Based on their experience and on a methodological tool they have developed, the authors identify four key ideas to keep in mind and six structuring questions to ask to support project leaders in preparing their participatory process.

In general, the first question that people ask themselves when they want to start a participatory process is: Where to begin? Many of the project leaders who we have supported wanted to set up a participatory process, either because they had followed a training course on a particular participatory method which they had enjoyed (forum theatre, role-play or other); or because they had had a successful "test experience" (a meeting with citizens, an online forum or other), which made them want to go further. Whether or not this is the case for you, we believe the first thing that is important to remember when embarking on a participatory process is to:

▶▶ Idea 1 – Think in terms of a process rather than a sequence of events

In both of the above cases, the leaders' attention is focused on a method (forum theatre, role-play) or on a specific participatory event (meeting, forum). These two elements are of course important, but there are other important questions to ask before proceeding.

Question 1: Why do you want to set up a participatory process? In other words, what is the objective of the participatory process?

Participatory methods and events are an actual means to an end. What is that end? Why do you want to involve different actors? The underlying question here is also: what do you want them to participate in?

In general, this chapter addresses decision-making. The decision may be simple (e.g. deciding whether to maintain or remove a retainer) or more complex and involve a range of actions and stakeholders (e.g. deciding how to control flooding in a territory).

Both the nature of the decision and the constraints linked to it (timetable, deadlines, budget, etc.) condition the participation methods that can be chosen. Whatever the decision, the main thing is to leave some room for manoeuvre for participation (see chapter 2).

⇥ Idea 2 – Leave room for participation in decision-making

Because if everything is already decided, what is the point of bringing people in to participate? At best you will create frustration, at worst a feeling of manipulation. We often hear statements from participants such as: "In the end, they only expected us to validate the principle", "Our opinion was not taken into account". The consequence? Distrust, even hostility towards the initiator of the process, rejection of the decision taken, and above all, the desire to never come back to participate, in other words virtually the exact opposite of what was intended. It is however possible to propose different levels of involvement in the decision (figure 9.1), depending on your objectives, your means as well as your constraints. What is important is that there is room for manoeuvre and that it is explained to all participants from the start of the participatory process (see chapter 4).

Once the objective of the process has been determined, it is time to look into the mechanics of participation, i.e. to "get your hands dirty". We deliberately use this technical metaphor, since the term generally used to describe this entire thought process is participation "engineering" (see box 9.1).

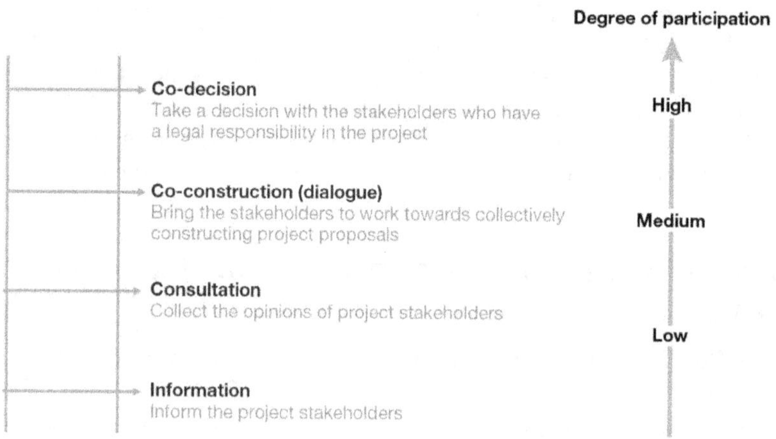

Figure 9.1. Participation scale (adapted from Arnstein, 1969; Lisode, 2017)

Based on a previously defined objective, the next step is to design a participation plan to achieve this objective (figure 9.2). Through a series of questions, the PrePar methodological approach, which stands for "preparing for participation", helps you construct such a plan. This approach was formalised by researchers from the G-EAU joint research unit "Water Matters" in Montpellier. It is part of the CoOPLAGE[1] approach presented in chapter 2.

---

1. Coupler des Outils Ouverts et Participatifs pour Laisser les Acteurs s'adapter pour la Gestion de l'Environnement = Coupling Open and Participatory Tools to Let Actors Adapt for Environmental Management.

> **Box 9.1. Participation engineering: definition and origins**
>
> Participation engineering can be defined as "a type of meta-level engineering and organisational decision-making that defines the rules and processes of collective choice in water management policy and planning" (Daniell *et al.*, 2010). In concrete terms, this engineering takes the form of a thought process to define the objectives, design, choice of methods, implementation, and monitoring and evaluation of a participatory process.
>
> The "participatory engineering of participation", also called co-engineering of participation, differs from the engineering of participation in that this reflection is carried out by a mixed group of actors, including future participants. The group may include the process initiator, the facilitator, elected officials, specialists and any other participant targeted by the participatory process.
>
> The term "participation engineering" comes from a view of engineering that applies not only to mechanical processes, but also to cognitive and decision-making processes (IEA, 2000; March, 1978). It also takes into account collective action and the social processes associated with practical engineering (Bucciarelli, 1994).

Question 2: Who should be involved?

A distinction should be made between those concerned (i.e. all the actors potentially affected by the decision or who can influence it) and those who may actually participate in the participatory process. First, draw up the most exhaustive list possible of all the stakeholders potentially affected by the decision in question: Who could be affected? Who could influence the decision? Who could be interested in the decision? Who could oppose it? Who could defend it? Then, decide which of these actors should become "participants" by choosing at which stage(s) each actor or category of actor should participate and in which capacity (see question 5).

⇉ Idea 3 – Consider all the stakeholders involved in water management (users, managers, etc.) and in participation (facilitator, lead, warrant, etc.)

There are various ways of developing a stakeholder map based on the interests of the different stakeholders, their power, their role in the decision, etc. (Hassenforder *et al.*, 2020, p 29-31). A fairly simple and pragmatic way of doing this is to consider broad categories of stakeholders and to list under each category the individuals and organisations implicated in the region. Figure 9.3 gives broad categories of actors often linked to socio-ecological sustainability which can be used as a guide. To ensure that no one is forgotten, the "snowball" technique used in social sciences can be quite effective. It involves asking the above questions (Who may be affected? Who may influence the decision? etc.) in regard to the stakeholders already listed to see if anyone has been forgotten.

In addition to the stakeholders involved in the decision, the list should not forget the actors whose role is dedicated to participation, such as those presented in table 9.1 and figure 9.4.

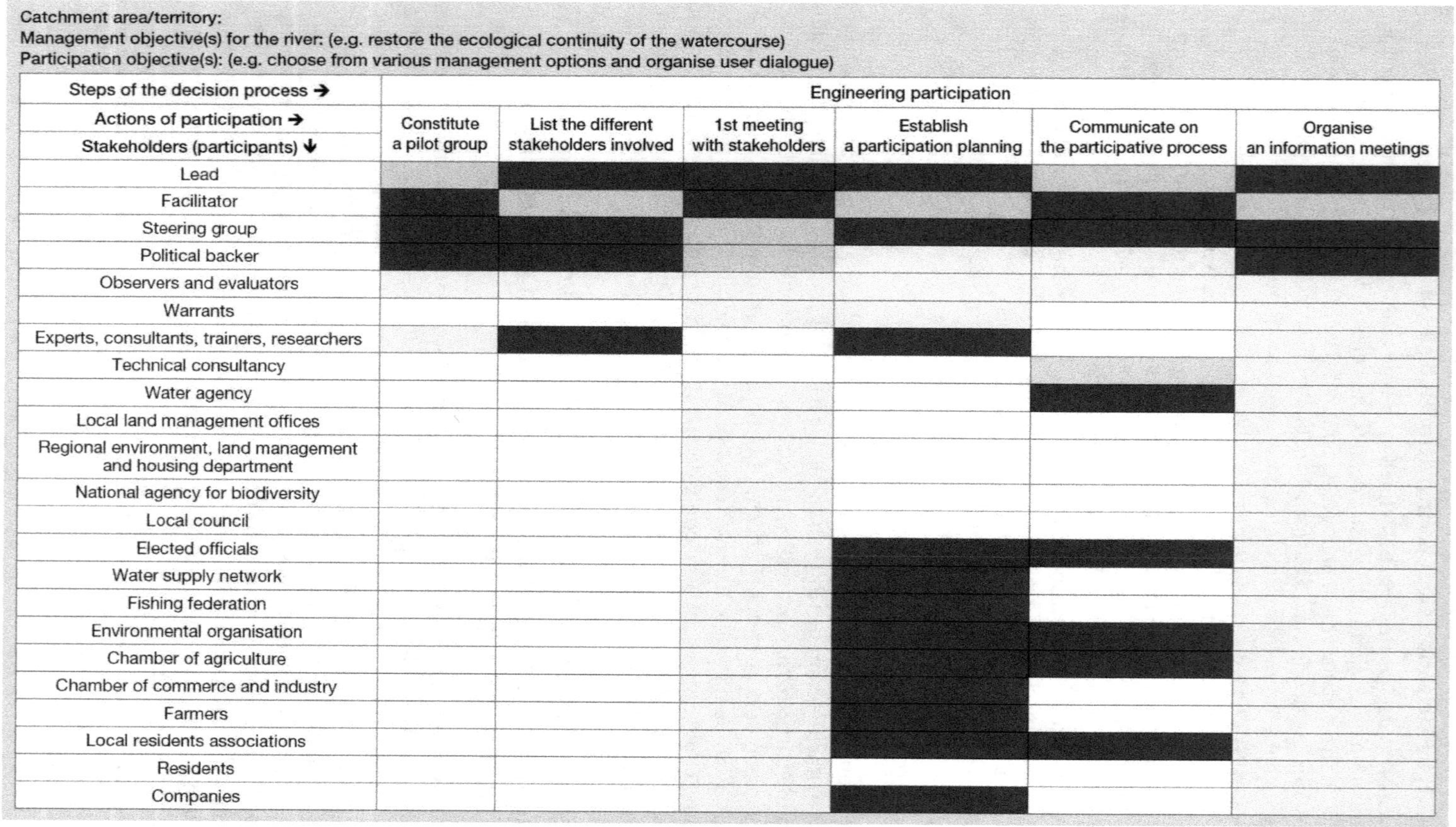

Catchment area/territory:
Management objective(s) for the river: (e.g. restore the ecological continuity of the watercourse)
Participation objective(s): (e.g. choose from various management options and organise user dialogue)

Steps of the decision process →	Engineering participation					
Actions of participation → Stakeholders (participants) ↓	Constitute a pilot group	List the different stakeholders involved	1st meeting with stakeholders	Establish a participation planning	Communicate on the participative process	Organise an information meetings
Lead						
Facilitator						
Steering group						
Political backer						
Observers and evaluators						
Warrants						
Experts, consultants, trainers, researchers						
Technical consultancy						
Water agency						
Local land management offices						
Regional environment, land management and housing department						
National agency for biodiversity						
Local council						
Elected officials						
Water supply network						
Fishing federation						
Environmental organisation						
Chamber of agriculture						
Chamber of commerce and industry						
Farmers						
Local residents associations						
Residents						
Companies						

Figure 9.2. Example of a participation plan made using the PrePar approach (information filled in only for step 1 "Engineering participation")

Colours correspond to the stakeholder's role in each of the participatory actions: green, organiser; black, active participant (provides opinions, decides); grey, passive participant (is present, listens, is informed); white, does not participate, is absent.

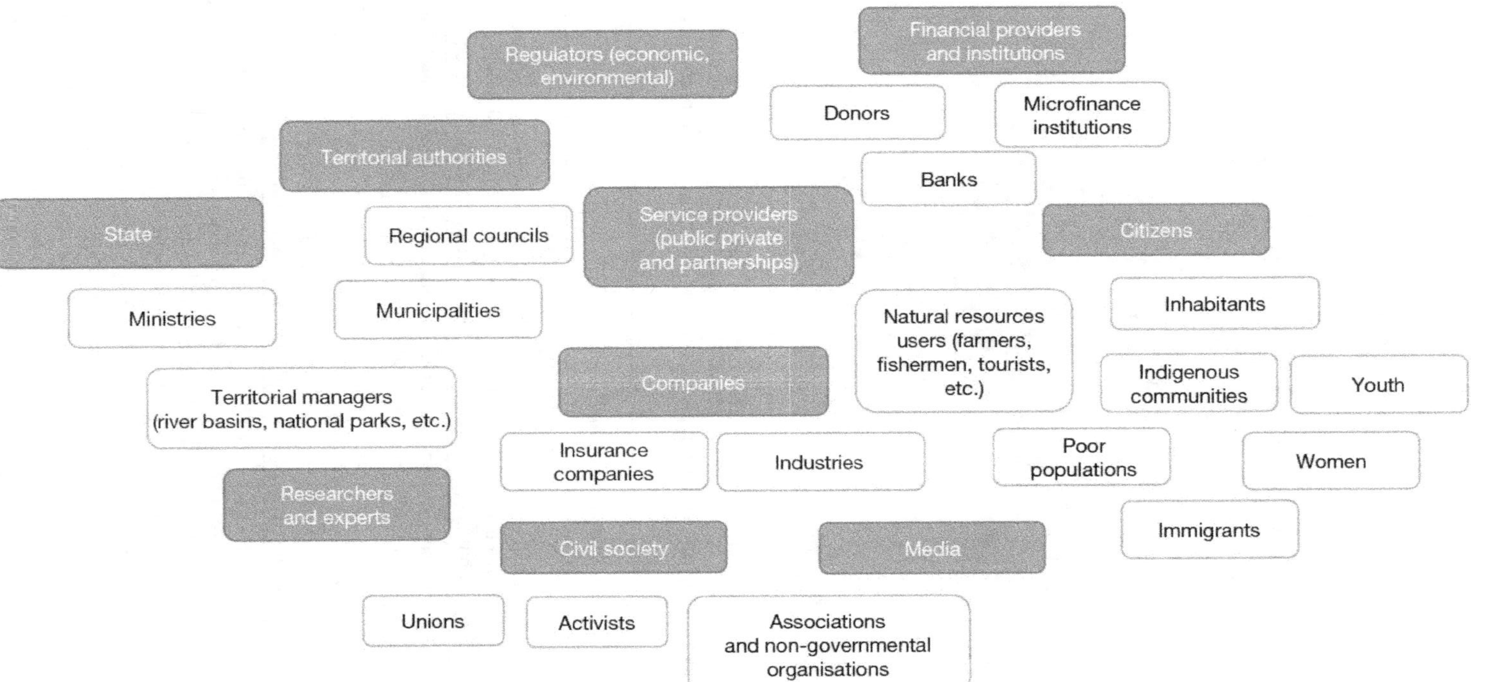

Figure 9.3. Broad categories of actors often linked to socio-ecological sustainability

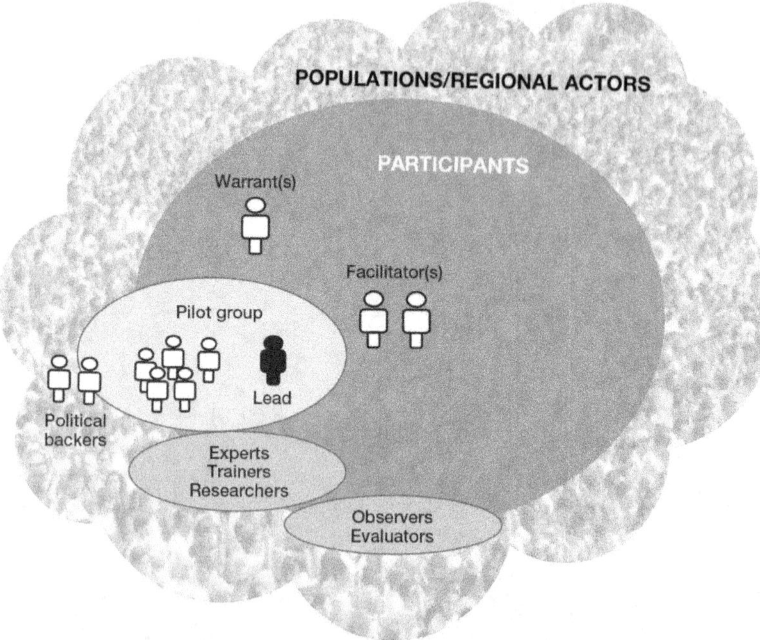

Figure 9.4. Actors dedicated to participation (definitions of the various roles are indicated in Table 9.1)

Table 9.1. Actors dedicated to participation: roles and definitions (Source: Ferrand *et al.*, 2017)

Actors dedicated to participation	Roles and definitions
Lead	Is the initiator of the participatory process. She/he ensures the operational and administrative management of the participatory process with the facilitator (organisation of events, mobilisation of participants, link between the different actors, etc.).
Facilitator	Is responsible for organising, leading and facilitating all local actions with the different stakeholders.
Pilot Group (optional)	Supports the lead in making strategic choices regarding the participatory process. Should help the lead understand and cover the different issues, connect with the relevant networks, and mobilise the participants. It does not decide on the participatory process, but advises and supports it.
Political backers	Support the lead with the political backing of the participatory process. Help institutionalise the participatory process and defend it with regard to elected officials and management bodies, and ensure that participation is given room for manoeuvre in decision-making.
Observers and evaluators	Contribute to the monitoring and evaluation of the participatory process and its effects by reflecting on the framework, collecting and/or analysing data, sharing results. They generally attend the various participatory events to draw up the attendance list, take notes on the discussions and contributions, distribute questionnaires if any, and write up a summary.

Actors dedicated to participation	Roles and definitions
Warrants	Ensure compliance with the rules and good conditions for participation (CNDP, 2023). See chapter 4.
Experts, consultants, trainers, researchers	Accompany the lead and the facilitator in the design, implementation and/or monitoring and evaluation of the participatory process. This support can take the form of training, advice, meetings or informal discussions.

Question 3: What are the steps?

The decision-making process, i.e. the different stages leading to a decision, can be broken down into different steps (figure 9.5). Several of these steps are fairly generic and are common to all decision-making processes: a diagnostic, also sometimes called an inventory, is often carried out whether it concerns the development of a Water Development and Management Plan (SAGE[2]), a Flood Prevention Action Programme (PAPI[3]), or a development project (e.g. construction of banks to combat erosion). A description of these different steps is available in the step's sheets presented in Irstea and AERMC (2016).

Depending on the decision-making process being considered, not all of these steps may be relevant. For example, the stage for scenario exploration or foresight may be relevant in the case of a Quantitative Water Resource Management Plan (PGRE[4]) to discuss different scenarios related to climate change or population growth and their impact on water availability and allocation of the resource between different uses. But this step may not, for instance, be relevant for a hydro-morphological restoration project.

These steps do not necessarily take place in the order shown in figure 9.5. Monitoring and evaluation, for example, takes place throughout the process and not just at the end (see chapter 10). A choice/priority/vote can be proposed to the participants in order to choose between different possible scenarios, and not necessarily after the identification of actions and plans. These steps are given as an indication to help you build a participation plan adapted to your situation. It is up to you to make them yours, to name and organise the steps so that they correspond to your project.

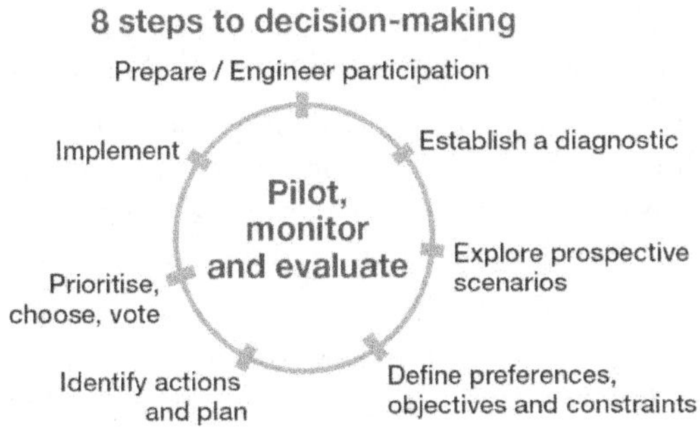

Figure 9.5. Generic steps in the decision-making process (source: Irstea and AERMC, 2016)

---

2. SAGE = Schéma d'aménagement et de gestion des eaux
3. PAPI = programme d'actions de prévention des inondations
4. PGRE = plan de gestion quantitative de la ressource en eau

Then, for each step, the desired degree of participation (low, medium or high, see figure 9.1) should be determined based on the descriptions provided in the step sheets (Irstea and AERMC, 2016).

Question 4: What actions should be taken?

For each stage, as in traditional project management, the next step is to list the actions to be carried out, i.e. detail the activities that will be conducted for each stage. For example, for the "structuring participation" stage, one might consider:
– establishing a pilot group,
– listing the different actors involved,
– meeting with the stakeholders to identify other potential participants and to present the proposed approach to them,
– establishing a participation plan,
– communicating on the participatory process (radio, flyers, digital displays, etc.),
– organising an information meeting,
– ...

These actions can be reported in the PrePar plan (figure 9.2).

Question 5: Who is involved in what steps and actions and in what capacity?

For each actor or group of actors, the objective is then to determine their role in each action (figure 9.6):

	Organizer	Organise, get things done
	Active	Give an opinion, decide (active participation)
	Passive	Be present, listen, be informed (passive participation)
	(Nothing)	Do not participate, be absent

Figure 9.6. Colors of boxes corresponding to roles played by actors in each of the actions

Green, organiser; black, active participant (provides opinions, decides); grey, passive participant (is present, listens, is informed); white, does not participate, is absent.

Figure 9.2 gives an overview of the participation plan obtained at the end of this step.

By going through the plan from top to bottom, you can then ask yourself whether for each stage the listed actors and their roles correspond to the expected degree of participation. For example, if you have selected a high degree of participation in the action proposal phase, does the plan actually foresee that most of the actors concerned will have an active role during this phase?

Reading the plan from left to right allows you to analyse at which stage(s) you plan to mobilise each of the different actors listed and to see if this mobilisation is consistent over time. For example, if you have planned to mobilise certain stakeholders only at the implementation phase, will they agree to implement a project on which they have not given their opinion beforehand? (the answer may be yes if it is a sub-contractor, for example, or no if they are citizens who are asked to reduce their water consumption without having been explained why).

Question 6: What participatory methods should be used?

The participatory methods listed in figure 9.7 and detailed in the method sheets (Irstea and AERMC, 2016) can help guide the choice of participatory methods at different stages.

Steps of the decision	Degree of participation	
	Co-construction	Co-decision
1. Prepare / Engineer participation	Participatory analysis by actors	
	Participation charter	
	Participation plan	
2. Establish a diagnostic	Participatory analysis by actors	Participatory photo & video
		Participatory modelling
	Participatory cartography	
	Participatory simulation/role-play	
		Participatory diagnostic
	Participatory theatre	
		Participatory observatory/ inventory/monitoring
	Summary of collective memory/participatory archive	
3. Explore prospective scenarios		Participatory scenarios/prospects
	Participatory simulation/role-play	
	Participatory theatre	
4. Define preferences, objectives and constraints	Preference elicitation	
	Citizens charter	
		Mental map
		Participatory planning
5. Identify actions and plan	Participatory map	
	Participatory budget	
	Participatory theatre	
		Mental map
6. Prioritise, choose, vote	Evaluation of actions and plans (per criterion)	
	Prioritisation and vote	
	Consensus building	
	Deliberation	
7. Implement	Participatory financing	Participatory observatory/ inventory/monitoring
	Participatory worksite	
	Participatory cleaning	
	Participatory monitoring and evaluation	
8. Pilot, monitor and evaluate		Participatory observatory/ inventory/monitoring
		Participatory photo & video

Figure 9.7. Examples of methods for co-constructing or co-deciding at each of the eight stages of the decision (Irstea and AERMC, 2016)

⏵ Idea 4 – Choose participatory methods according to the objectives, not the other way around

This list is not exhaustive. More transversal methods can also be used. They are not necessarily specific to one or more stages of the decision-making process (wish tree, brainstorming, World Café, focus group, etc.). Digital tools are also an integral part of these participatory methods. This is evidenced by the multiplication of private service providers and technological providers of "civic-tech" (civic technologies).

The French Etalab website (www.consultation.etalab.gouv.fr/) lists a certain number of open online consultation tools (see also Aucante *et al.*, 2020).

Table 9.2 summarises the six phases for designing a participation plan following the PrePar approach.

Table 9.2. The six phases for designing a participation plan (PrePar)

PrePar phases	Description
1. Formalise the objectives of participation	*Question 1: Why do you want to set up a participatory process? In other words, what is the objective of the participatory process?* Define the objectives; this can be done by the project leader alone (future pilot), or in discussion with the stakeholders
2. Identify stakeholders (participants)	*Question 2. Who should be involved?* Make a map of stakeholders. In addition to the water management stakeholders (elected officials, industries, associations, users, etc.), also consider the participation actors (facilitator, warrant, evaluator, etc.).
3. Validate the steps of the decision	*Question 3. What are the steps?* Using the step-by-step sheets, validate the order of the decision-making steps most relevant to the local participatory process and define the desired degree of participation. Eight decision-making stages can be mobilised: – Structure participation – Establish a diagnostic – Explore scenarios – Define objectives, preferences and constraints – Identify actions and plans – Choose, prioritise, vote – Implement – Monitor and evaluate
4. List the actions to be taken	*Question 4. What actions should be taken?* For each step, list the activities that will be needed to achieve the objectives.
5. Define the role of the actors for each action	*Question 5: Who is involved in what steps and actions and in what capacity?* Define the role of each actor for each action; these can be: – Organiser (O) = Organise, get things done – Active (A) = Give opinions, decide (active participation) – Passive (P) = Be present, listen, be informed (passive participation) – (Nothing) = do not participate, be absent

PrePar phases	Description
6. Discuss participatory methods	*Question 6: What participatory methods should be used?* For each activity, depending on the level of participation and the target audience, and on the resources available to you (financial and human resources, time and skills), identify the participatory methods to be used. Think about diversifying these methods and do not hesitate to go beyond what you usually do (through training for example). The choice of methods can be decided along with the actors involved and available skills can also be mobilised for their implementation.

▸▸ Conclusion

The particularity of participation engineering is placing the identification of stakeholders and their roles at the heart of the organisation and decision-making processes for water management planning. The PrePar method proposes a way of preparing and thinking about this engineering, but many others exist (e.g. Lisode, 2017; Graine Guyane, 2017; World Bank, 1996; OECD, 2015). The preparation of a participatory process can itself be participatory, i.e. involving the stakeholders who are concerned by the project. The advantages of this approach include a better appropriation of the objectives, greater adaptation of activities to the specificities of the field, and stronger commitment to the implementation of the approach. However, such co-engineering of participation itself requires preparation and dedicated resources, which should not be underestimated and thus risk creating disengagement.

In the course of our experiences, we have observed the importance of thinking about participatory ambitions in relation to the means available, and of being as explicit as possible with the actors concerned about the room for manoeuvre that will be allocated to them, as well as about the way in which the results of the participatory process will be integrated into the decision-making processes. In short, rather than multiplying participatory activities, it is better to focus on a few well-thought and prepared activities as a process to achieve a clearly formalised objective.

Box 9.2. Participation engineering in the Drôme

In preparation of the revision of the Drôme Water Development and Management Plan (SAGE), the Drôme River Joint Syndicate (SMRD*) decided to collect public opinions and proposals for action on the river and its management which were to be taken into consideration during revision of the SAGE. The originality of the approach was to involve the participants in the design, implementation and monitoring-evaluation of the participatory process itself (see insert 3 in chapter 17).

The co-engineering stage of participation took place from December 2016 to May 2017.

A group of 46 people, mostly citizens living in the Drôme catchment area, thus carried out the engineering of the participatory process through three successive one or two-day workshops over a period of six months in December 2016, February and March 2017. These workshops were led by a facilitator. They alternated between plenary sessions, group work and individual reflections, based on the steps presented in figure 9.8. A participation plan was thus co-constructed and implemented in 2017-2018.

Box 9.2. (next)

At the end of the participatory process, the group of participants submitted a citizen's diagnostic of the river to the Local Water Commission. It included 630 contributions, 189 proposals for action to improve management of the river, as well as a final report and five thematic summaries (Hassenforder *et al.*, 2020, 2021). These results were integrated into the subsequent revision of the SAGE.

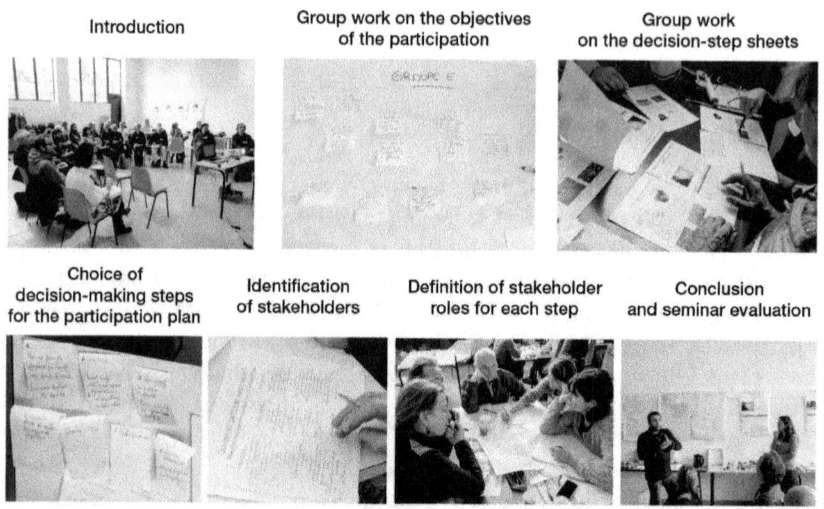

Figure 9.8. Participatory workshops: plenary sessions, group work, and individual reflection (© S. Girard and E. Hassenforder)

* SMRD = syndicat mixte de la rivière Drôme

▶▶ References

Arnstein S. R., 1969. A Ladder Of Citizen Participation. *Journal of the American Institute of Planners*, *35*(4), 216-224. https://doi.org/10.1080/01944366908977225

Aucante M., Hassenforder E., Ferrand N., Pressurot A., Sureau-Blanchet N., 2020. *Quelle stratégie participative pour la gestion locale de l'eau avec les citoyens ? Participation et numérique.* https://www.eaurmc.fr/jcms/pro_100740/fr/fiches-participation-numerique-irstea-aermc-10-06-20

Bucciarelli L. L., 1994. *Designing engineers.* MIT Press.

CNDP (Commission Nationale du Débat Public), 2023. *Garant.e.s de concertation. Notice d'information jointe à l'appel à candidatures.* https://www.debatpublic.fr/sites/default/files/2023-07/Notice_d-information_role_de_garant-2023.pdf

Daniell K. A., White I. M., Ferrand N., Ribarova I., Coad P., Rougier J. E., *et al.*, 2010. Co-engineering participatory water management processes: Theory and insights from Australian and Bulgarian interventions. *Ecology and Society, 15*(4).

Ferrand N., Noury B., Girard S., Hassenforder E., 2017. *Initial Guidelines on Stakeholders' Engagement and Year 1 Participatory Process in the PCS* SPARE WPT1 D.T. 1.1.2 Final report. https://spare.boku.ac.at/index.php/en/about-spare/initial-guidelines-on-stakeholders-engagement-and-year-1-participatory-process-in-the-pcs

Graine Guyane, 2017. Préparer un projet participatif #2, 8 fiches méthodologiques Avec témoignages et astuces des porteurs de projets rencontrés, Cayenne, https://graineguyane.org/wp-content/uploads/2017/09/Partie_2-Preparer-un-projet-participatif-1.pdf

Hassenforder E., Girard S., Ferrand N., Petitjean C., Fermond C., 2021. La co-ingénierie de la participation : une expérience citoyenne sur la rivière Drôme. *Natures Sciences Sociétés*, 2, 159-173. https://doi.org/https://doi.org/10.1051/nss/2021050

Hassenforder E., Pressurot A., Ferrand N., Aucante M., Sureau-Blanchet N., 2020. *Quelle stratégie participative pour la gestion locale de l'eau avec les citoyens ? Retours d'expérience et questions à se poser.* https://hal.archives-ouvertes.fr/hal-02976121

IEA (Institution of Engineers Australia), 2000. *Code of ethics.* Barton, Australia https://fr.scribd.com/document/464903885/Code-of-Ethics-2000.

Irstea, AERMC, 2016. *Quelle stratégie participative pour la gestion locale de l'eau avec les citoyens ? - État de la connaissance, fiches étapes et fiches méthodes.* https://www.sauvonsleau.fr/jcms/e_17247/quelle-strategie-participative-pour-la-gestion-locale-de-l-eau-avec-les-citoyens-

Lisode, 2017. *Guide de concertation territoriale et de facilitation.* http://www.lisode.com/wp-content/uploads/2017/02/Lisode_Guide_concertation.pdf

March J. G., 1978. Bounded rationality, ambiguity, and the engineering of choice. *The Bell Journal of Economics*, 9, 587-608.

OECD (Organisation for Economic Co-operation and Development), 2015. *Stakeholder Engagement for Inclusive Water Governance.*

World Bank, 1996. *The World Bank Participation Source Book.* Washington, D.C., 280 p., http://documents1.worldbank.org/curated/en/289471468741587739/pdf/multi-page.pdf

Chapter 10

Evaluating a participatory process

Emeline Hassenforder and Nils Ferrand

One can argue that *"Setting up a participatory process is already cumbersome, adding an evaluation on top of that is beyond any ambitions I may have".* And yet... What's the point of involving different stakeholders, if in the end, you cannot tell if that participation has truly served a purpose? Following this statement, we invite you to read this chapter. We assert that evaluation is not synonymous with depression. Your evaluation can be adapted to suit your ambitions. Let's get to it!

▸▸ The ABCs

The ABCs of evaluating a participatory process are to be able to say how many people took part in the process, whether there were more women than men, whether there were only environmentalists and no representatives from the agricultural sector, or whether an elected official monopolised the floor and invited participants only had five minutes to express themselves.

▸▸ How do you go about it?

Assess the participants' demographics

First, ask yourself what you want to know about the participants. For example, if you want to know if the participants are representative of the region's population, ask yourself what data you have on the territory's population that you can compare with the data you will collect from your participants. If you are going to use national statistics, you can use the same indicators as those used by the national bureau for statistics. This will make it easier for you to compare data later on. The same applies to gender, place of residence, socio-professional category, etc. This allows you to establish a list of individual characteristics and associated options that you want to know about the participants.

Example of characteristics: age, gender, place of residence, socio-professional category, type(s) of river use(s), household composition, community or volunteer activities, telephone number or email, etc.

Example of associated options: age >18/19-24/25-64/65-79/>80 years old.

There are several means for collecting this information, each with its advantages and disadvantages (table 10.1). This table is of course non-exhaustive.

Table10.1. Possible means for collecting individual characteristics of participants

	Advantage	Disadvantage
Online pre-registration	Automatic data collection and processing, provides information on those who will be present	May inhibit participation of some people who just want to "come and see"
Registration on arrival at the 1st participatory event	Allows organisers to immediately see who the participants are	Requires the support of an organiser to ensure that everyone has registered, and has entered their data
Pass around an attendance list requesting information on the individual characteristics of participants	Simple to set up and customary for most participants	Some people do not wish to share personal information such as their age or residence with other participants

What about anonymity?

It is, of course, possible to organise a participatory process where everyone remains anonymous. This is the case with most public meetings, where no registration is required and everyone can participate and speak without even having to introduce themselves. Again, there are advantages and disadvantages to this option. On the one hand, this helps limit prejudices between participants (*"He's eco-friendly"*, *"She's a right-wing mayor"*). On the other hand, if a decision is made, you will not be able to justify that the room was not in fact filled with members from the National Federation of Farmers' Unions or from environmental activists who came to sway the decision in their favour. There are several options in-between absolute anonymity (nobody knows who is who) and extensive demographic analysis. It is possible, for example, for the participatory process organiser to collect data on the participants, to present them with generic results (percentage of representatives from civil society, percentage of representatives from the administration, etc.) while maintaining individual identities anonymous.

The European General Data Protection Regulation (European Union, 2016) provides principles and steps to be followed when a public or private organisation collects and processes personal data. This includes informing participants about the type of data collected, for what purpose, by whom, who has access to the data and to whom it will be communicated, data retention periods, etc.

▸▸ Monitoring and evaluating the process

At the very least, information on who participated in which participatory event(s) is necessary. To do this, you can simply pass around an attendance sheet as mentioned above, or ask participants to pre-register or register upon arrival. The individual characteristics mentioned above are only collected once at the beginning of the process. At subsequent events, only the person's first name and surname or participant number (if you have chosen to assign a number to each person) will be requested. This information can then be entered into an Excel file (one row per participant, one column per event, and in each box a "1" if the person participated in the event, if not then nothing). This allows for a quick analysis of the number of participants at each event, the retention rate (did participants who came to the first event come back again?)

and the composition of the group of participants at each event. For process facilitators, these data are essential to adapting the participatory process along the way. For example, it allows you to determine whether it is better to organise an event in the evening or during the day depending on the targeted participants, whether the events upstream of the catchment area have attracted a particular socio-professional category and those downstream another, etc.

Above and beyond data on the number and characteristics of participants and events, you can also monitor and evaluate the progress of the process itself, for example:
– whether all participants were able to express themselves;
– whether the necessary documents were made available to the participants;
– whether the facilitator distributed speaking time in a balanced manner;
– whether tensions or conflicts emerged between participants;

– ...

There are various reference systems that propose "standard elements" to be evaluated in order to determine whether or not a participatory process is going well. For example, in the insert following this chapter, there is a focus on the Participation compass developed by Cerema[1]. This compass is based on the values and principles defined in the participation charter of the French Ministry for the Environment, Energy and the Sea[2] (see also chapter 4). Other guidelines exist that define the principles of "good" participation. The ones best known in the field of participation research are those by Gene Rowe and Lynn J. Frewer (2000), which include nine acceptance and process criteria:

Acceptance criteria:	Process criteria:
– representativeness of the participants,	– accessibility to resources,
– independence of the participants,	– definition of roles for each participant,
– early involvement,	– structure and clarity of decision-making,
– influence on final policy,	– cost effectiveness.
– transparency of the process.	

The moderators of a participatory process can use an existing reference framework to evaluate the progress of their process. They can also define the criteria themselves to include those that seem most relevant to them for evaluating the effective progress of their process. It may also be pertinent to involve a small group of five to ten people to reflect on this, each with a different point of view on what constitutes a "good" participatory process. This monitoring and evaluation steering group can further help to ensure that these pre-defined principles of good participation are respected throughout the process. They can also contribute to the collection or analysis of data and the sharing of results.

These principles of good participation often constitute the content of participation charters, which are communicated to and endorsed by all participants. Monitoring and evaluation therefore directly supports the implementation of the participatory process.

In addition, the participatory process facilitators can call on one or more participation warrants, whose role is precisely that of independently ensuring the rules of

---

1. Centre d'études et d'expertise sur les risques, l'environnement, la mobilité et l'aménagement (Cerema): Centre for Studies and Expertise on Risks, the Environment, Mobility and Urban Planning is a French public agency for developing public expertise in the fields of urban planning, regional cohesion and ecological and energy transition.
2. https://www.ecologie.gouv.fr/sites/default/files/Charte_participation_public.pdf

participation are respected, in compliance with general principles or a local charter. The warrant may be a local person, who then withdraws from the participatory process in order to remain neutral. There are also professional warrants who may have an official role in the procedures monitored by a national organisation such as the National Commission for Public Debate in France.

This is a two-fold process in which the principles of good participation are defined followed by the collection of data on whether or not these principles have been respected. This is what a warrant or participation observer is supposed to do. For instance, if one of the principles is that everyone should have the opportunity to speak, the observer will note who spoke at the various events, possibly for how long, and whether the facilitator offered the possibility to speak to those who had not yet spoken.

▸▸ Assessing the impacts of the process

In the previous two paragraphs, we discussed procedural evaluation, i.e. evaluation of the process as such, as opposed to impact evaluation, which aims at measuring the effects of the process on the participants (e.g. Did they learn something?), on the project or policy (e.g. Was the project modified following proposals made by citizens?) or on the initiator of the process itself (e.g. Is the water manager implementing participatory processes in a more systematic manner following this process?).

What is important here is the impact you want to achieve with your process on your territory. This is what needs to be assessed. Keep in mind that different stakeholders may have different visions of the impact expected from the process. This is why we advise you to carry out the following steps with a small group of people who will be in charge of monitoring and evaluation (table10.2).

Table 10.2. Steps to developing a monitoring and evaluation protocol
(source: Hassenforder *et al.*, 2016; Hassenforder *et al.*, 2018)

Steps	Questions to ask	Example
Identify the objectives of the evaluation	What are the impacts we want to assess?	We want to assess whether the participants have learned something during the process.
Define the indicators	What do we need to know to be able to assess these impacts?	We want to know if the participants learned how their watershed works from a hydrological standpoint.
Check feasibility	Will we be able to collect and analyse data on the listed indicators?	Will participants be willing to answer questions about their knowledge? Is there sufficient budget for collecting and analysing this data? Is it really useful? To whom? Etc.
Identify monitoring and evaluation methods	By what means will we collect this data (questionnaires, interviews, surveys, observation of participatory events, photos, videos, etc.)?	Questionnaire: ask participants at the end of an event if they have learned how their watershed works from a hydrological standpoint. Observation: note what participants say about the catchment area (e.g. "I didn't know that my tap water came from aquifer X"). Mapping: ask participants to draw the catchment area before the start of the process and at the end of the process.

Steps	Questions to ask	Example
Implement the evaluation	Who will collect the data using these monitoring and evaluation methods, when and with what resources (budget, time)?	An evaluator has been hired to observe the participatory events and record the content of the exchanges. Data analysis is done by a researcher.
Analyse the data	What do the data say about the impacts initially identified? Are there any unexpected effects?	23 out of 34 participants said they have learned something about the hydrological functioning of the catchment area. Of these 23 people, 19 thought that their tap water came from the river. The workshop provoked a debate on the transition to private sector management of the drinking water supply in the municipality of XX.
Share results	With whom do we want to share the results and how (written reports, press articles, videos, oral presentations, etc.)?	An infographic will be posted on the district's website and sent to all participants by email. An in-depth analysis will result in a scientific paper. A press article in the local newspaper will mention the main results.

Several types of impacts can be generated (and evaluated) by participatory processes. Table 10.3 lists some of these as a guide and figure 10.1 shows some examples of simple monitoring and evaluation methods.

Table 10.3. Types of impacts that can be generated (and evaluated) by participatory processes (Source: Ferrand and Daniell, 2006)

Type of impact	Explanation	Possible monitoring and evaluation methods
External (E)	Environmental, economic, social, cultural, political or institutional impacts	Environmental impact study, cost-benefit analysis, etc.
Normative (N)	Impacts on the norms, values, preferences, goals of participants: e.g. whether they favour the short or long term, conservation or innovation, cooperative or individual, etc.	Questionnaire, cognitive mapping, simulations, etc.
Cognitive (C)	Impacts on representations, beliefs and/or knowledge about the project, the environment, the social framework, others, solutions	Questionnaire, cognitive mapping, simulations, etc.
Operational (O)	Impacts on the practices, actions and behaviours of actors	Direct observation, direct or indirect reporting, external evaluation, etc.
Relational (R)	Impacts on relationships between participants: e.g. trust, solidarity, mutual understanding, tensions, conflicts	Mapping of actors: powers, interests, social networks, political networks, etc.

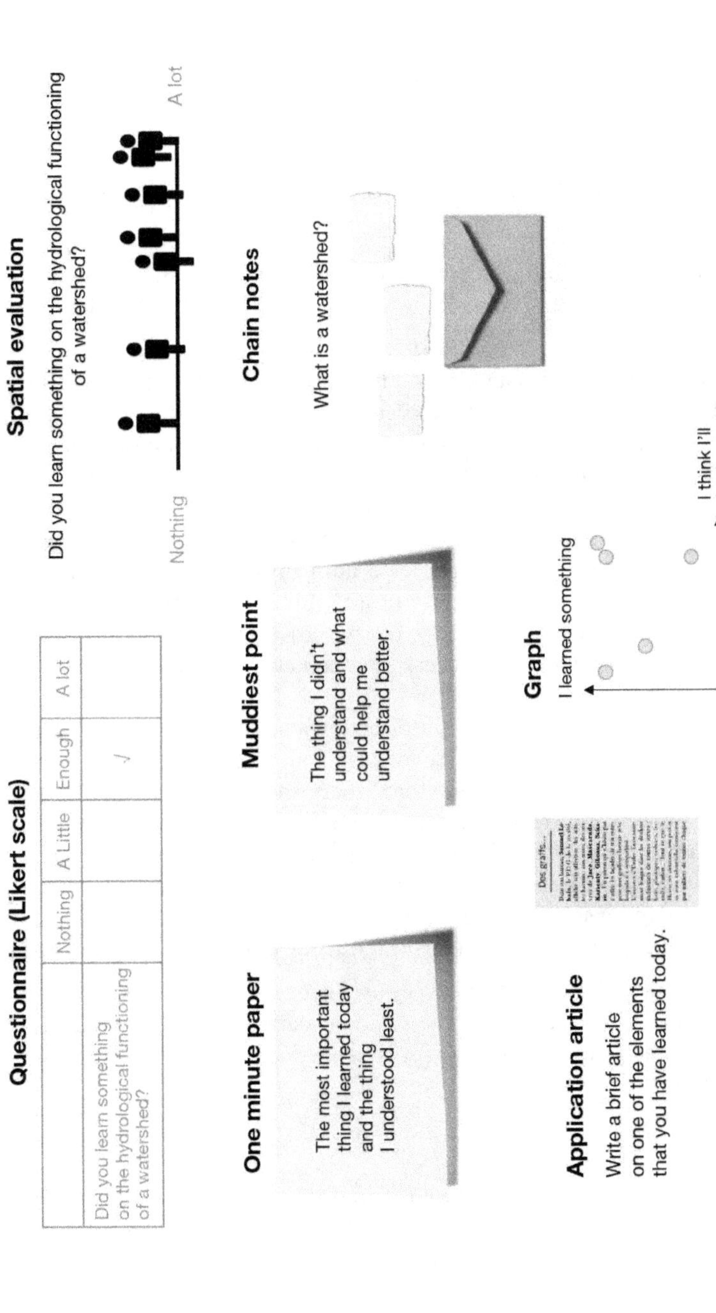

Figure 10.1. Examples of easy-to-implement monitoring and evaluation methods

Minute paper, muddiest point, chain notes and application article are from Classroom Assessment Techniques (CATs), which are short 5-20 minute activities to assess learning (Angelo and Cross, 1993; Bachy and Lebrun 2009; Univ. Iowa, 2020; Univ. Vanderbilt, 2020)

Type of impact	Explanation	Possible monitoring and evaluation methods
Equity (E)	Impacts on the distribution of material and immaterial resources among the actors mobilised in the project: e.g. knowledge, influence, control, risk, etc.	Simulation, questionnaires, interviews, JustAGrid[a] (allocation game on social justice principles)

[a] http://cooplage.org/tools/just-a-grid

⇥ Do not wait until you have finished the participatory process to evaluate it!

Evaluation is all the more useful when it is done along the way. Why wait until the end of the process and produce a nice report that no one will read? (A little cynicism never hurts!).

Evaluating as you go, or *in itinere* evaluation for those in the know, allows for you to:
– find out whether the participating audience is indeed the target audience. For example, if the process is aimed at young people and the evaluation shows that the majority of participants are in their fifties, this assessment will allow you to adapt your process to try to reach young people more effectively, e.g. by using online social networks, by including workshops in schools for older students, etc. Now, you will tell me that the facilitator will have noticed if the participants are more wrinkled than spry. Indeed, but what about an online participatory process? And what if this data could be used to enhance the process and attract more people? For example: 250 youth from your city have already taken part, what about you? Your opinion counts as well!
– know if the process is going in the right direction and has the expected effects. For example, in the scope of a water resource management plan that is set up to improve the sharing of water resources in a territory where there is a shortage, an ongoing evaluation may allow you to realise that the local population think that the farmers consume the most water, where in fact the majority of water consumption is domestic. Knowing this will allow for it to be discussed, for figures to be put on the table, and for informed solutions to be sought. Without the assessment, the locals would probably have proposed an array of solutions aimed at reducing agricultural consumption.

Moreover, reflection on the evaluation is very useful for reflecting on the process itself. As mentioned above, thinking about what a "good" participatory process is from the standpoint of the different actors involved, is as useful for the evaluation as for the construction of the process itself.

⇥ Who evaluates?

Different people can contribute at different stages of monitoring and evaluation. The reflection on objectives and indicators, for example, can be done by a small group of five to ten people dedicated to monitoring-evaluation, and then one or more external people can be hired to collect and analyse the data. The initiators of the participatory process can also choose to evaluate themselves and/or ask the participants to do so. In most cases, monitoring and evaluation is carried out by a number of actors. This allows for a division of labour and the valuing of multiple viewpoints. Whatever choices you make, each has advantages and limitations. For example, hiring an outsider can bring a "fresh" perspective to the process, but participants may be more reluctant to confide in someone they do not know.

No matter who evaluates the participatory process, we consider that monitoring and evaluation is always subjective. Even if an external person is brought in, this person, because of age, gender, employer, geographical origin and own knowledge, will have a certain view of the process. The people they survey to collect data (participants, organisers) will also have their own point of view on this person, which will at least partially condition their answers. This subjectivity is an integral part of monitoring and evaluation. The trick is to turn it into an advantage rather than an obstacle and to take it into consideration when defining who is evaluating.

▸▸ Conclusion

We hope that we have convinced you that the evaluation of a participatory process can be integrated into the participatory process itself. Evaluation guides you into asking the right questions when developing the participatory process, putting the multiplicity of viewpoints and expectations up for discussion from the outset, and avoiding possible conflicts and disappointments at a later stage. The evaluation also allows you to adapt the process along the way, for example if the participants are not those expected or if the proposed subject of debate does not respond to the issues that concern the majority of the actors in the field. Finally, the results and impacts of the process can be highlighted and supported on the basis of concrete data, as an evidence-based study.

The evaluation of a participatory process is not insurmountable; it is not reserved for scientists or experts. It is within the reach of anyone who takes the initiative to do it and can be adapted to the ambitions and resources that are available. It is entirely possible to design and implement the monitoring-evaluation of a process from start to finish; it is just as possible to outsource part of it or to rely on existing guides and protocols (e.g. the Cerema compass—see insert at the end if this chapter, the ENCORE approach proposed in table 2.1, Rowe and Frewer's evaluation criteria 2000; Daré *et al.*, 2020; or other approaches presented in Concertation décision environnement, 2009). The only thing to remember is to be able to answer the questions you ask yourself, and to remain open to the surprises and unexpected effects that any participatory process may generate.

▸▸ References

Angelo T.A., Cross K.P., 1993. *Classroom Assessment Technologies*, Second Edition, San Francisco: Jossey-Bass Publishers.

Bachy S., Lebrun M., 2009. Catégorisation de techniques de rétroaction pour l'enseignement universitaire, *Mesure et évaluation en éducation*, vol. 32, n° 2, p. 29-47. https://doi.org/10.7202/1024953ar

Concertation décision environnement, 2009. *Qui est vraiment prêt pour évaluer la concertation ?*, Compte-rendu du séminaire Concertation Décision Environnement du 10 juin 2009, www.concertation-environnement.fr/documents/seminaires/CDE_Seminaire_Permanent_1.pdf

Daré W., Hassenforder E., Dray A., 2020. *Observation manual for collective serious games*, CIRAD, Montpellier, 68 p. https://doi.org/10.19182/agritrop/00144

European Union, 2016. Regulation (EU) 2016/679 of the European Parliament and of the Council of 27 April 2016 on the protection of natural persons with regard to the processing of personal data and on the free movement of such data, and repealing Directive 95/46/EC (General Data Protection Regulation) https://eur-lex.europa.eu/eli/reg/2016/679/oj

Ferrand N., Daniell K.A., 2006. *Comment évaluer la contribution de la modélisation participative au développement durable ?*, Séminaire interdisciplinaire sur le développement durable, 30 novembre 2006, Lille, France, 20 p. https://halshs.archives-ouvertes.fr/halshs-02933305

Hassenforder E., Ferrand N., Girard S., 2018. *Guideline on monitoring and evaluation methods for "Local Capacity in River Protection and Management",* SPARE WPT1 Deliverable 1.3.1 Report, Montpellier, https://spare.boku.ac.at/media/other/MandE_Guidlines.pdf

Hassenforder E., Pittock J., Barreteau O., Daniell K.A., Ferrand N., 2016. The MEPPP framework: A framework for monitoring and evaluating participatory planning processes, *Environmental Management Journal*, 57, p. 79-96. https://doi.org/10.1007/s00267-015-0599-5

Rowe G., Frewer L.J., 2000. Public Participation Methods: A Framework for Evaluation, *Science, Technology, & Human Values*, vol. 25, n° 1, p. 3-29. https://doi.org/10.1177/016224390002500101

Univ. Iowa, 2020. Classroom Assessment Techniques (CATs), https://www.celt.iastate.edu/teaching/assessment-and-evaluation/classroom-assessment-techniques-quick-strategies-to-check-student-learning-in-class/

Univ. Vanderbilt, 2020. Classroom Assessment Techniques (CATs), https://cft.vanderbilt.edu/guides-sub-pages/cats/

"Participation compass", a web application to organise and track participatory processes

Anne Hilleret and Karine Lancement

The "Participation compass" is a web application designed to guide in the development of a participatory approach. It provides indicators that can be shared, or even developed jointly with all the participants. It can also be used to assess completed projects and generate a report. Initially designed for use by local authorities, the Participation compass can be used by anyone who is involved in any role of a participatory process.

The Participation compass is a tool developed by Cerema[1]. This tool is intended to support and guide project leaders (local authorities, contracting authority, citizen groups, etc.) in defining, implementing step-by-step and self-assessing participatory processes with regard to the principles and values set out in the Public participation charter developed by the French Ministry for the Environment, Energy and the Sea[2] (see also chapter 4).

The aim of the Participation compass is to guide the project leader or group in:
– defining the ambition and preparing the participatory process,
– following-up the process,
– drafting a final report and qualitative evaluation of the process with regard to the principles and values of the Public participation charter.

The compass is intended for any project leader or collective, including those who are new to the field of participation. The tool is nevertheless easier to use with a minimum of training in the field. It is intended to be used as a dialogue tool for project partners and elected officials.

The Participation compass was developed primarily for inter-municipality projects managed by local authorities. However, it can be adapted to any kind of participatory project or process, whether voluntary or regulatory and on any scale. Projects, instances and processes that can benefit from this tool include:
– Agenda 2030, regional agri-food projects;
– Territorial climate-air-energy plans (PCAET), territorial coherence plans (SCOT), local urban plans (PLU)[3];
– Urban regeneration projects, eco-district projects;
– Forward planning;

---

1. Centre d'études et d'expertise sur les risques, l'environnement, la mobilité et l'aménagement (Cerema): Centre for Studies and Expertise on Risks, the Environment, Mobility and Urban Planning is a French public agency for developing public expertise in the fields of urban planning, regional cohesion and ecological and energy transition.
2. https://www.ecologie.gouv.fr/sites/default/files/Charte_participation_public.pdf
3. plan climat-air-énergie territorial; schéma de cohérence territoriale; plan local d'urbanisme

– Specialist or experimental participatory methods (Spiral, Visions+21, etc.);
– Participatory bodies (development committees, regional economic, social and environmental councils, citizen assemblies, etc.).

When a local authority is conducting several participatory processes at the same time, using the compass for each project helps to provide an overview of all the processes (timetable, types of audiences targeted). It can also be used to share and capitalise on the tools implemented.

▸▸ A four-step process

Using the Participation compass is very simple:
– request access to the application by contacting Cerema at the following address: boussole-participation@cerema.fr
– download the two blank files from the two corresponding tabs as well as the instructions:
– "self-assessment" file
– "step-by-step" file
– user manual;
– browse the manual and annexes;
– fill in the two files; this is the core of the work.
– self-assessment will be all the more relevant if it is carried out by a group of actors and shared with all project stakeholders;
– the "step-by-step" contains elements to formalise the project, monitor the mechanisms, communicate on and account for the process; they are mentioned at the end of this chapter;
– project formalisation will be all the more relevant if it is carried out by a group of actors and shared with all project stakeholders;
– project monitoring: there are useful documents in the user manual annexes (list of participants, evaluation, etc.); useful tip: enter information as you go along so as not to have too much information to enter at once;
– upload the completed files into the corresponding tabs in the application: "self-assessment" and "step-by-step":
– the application works even if all the data are not filled in; if there is no data, a message to this effect is displayed;
– the application presents the results in the form of clickable graphs (successive clicks provide access to different types of information); these graphs can be exported as images (.png) so that they can be inserted into communication or reporting documents (assessment report, etc.).
– caution: the application formats results but does not interpret the information.

▸▸ Two possible uses

There are two proposed sections in the compass referring to two possible uses:

Section 1 – Self-assessment

As a self-implemented method of evaluation, self-assessment can be used upstream of the participatory process to define ambitions in terms of participation as well as downstream to evaluate them. Self-assessment is based on the elements in the Public participation charter proposed by the Ministry for the Environment, Energy and the Sea (see chapter 4). For each element and sub-element, each person can evaluate their participatory approach with regard to the ambition in the charter, on a scale from one to five (figure I1.1), from "outside the charter" (level 1) to "becoming exemplary and innovative" (level 5). An example is presented in table I1.1.

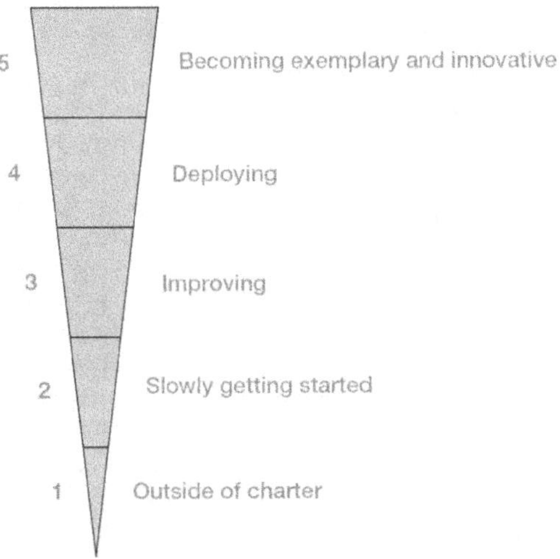

Figure I1.1. Each element of the Public participation charter is assessed on a one-to-five scale

Each sub-element of the charter is thus defined for each level, which allows the users to position themselves for each element. The results of this self-assessment can be visualised in the form of a circular diagram (figure I1.2).

Figure I1.2. Example of a circular diagram representing self-assessment results
(Source: Cerema)

Table I1.1. Example of a participatory process self-assessment with regard to the Participation Charter proposed by the Ministry for the Environment, Energy and the Sea

	Are the nature of the project and its stakes, as well as the need it addresses, clearly presented?				
Level	**1** **Outside of charter**	**2** **Slowly getting started**	**3** **Improving**	**4** **Deploying**	**5** **Becoming exemplary and innovative**
Definition	No written records	Nature of the project and the issues at stake drawn up and presented internally	Nature of the project and the issues at stake drawn up internally and presented to stakeholders	The nature of the project and the issues at stake drawn up internally and clearly presented to stakeholders (with particular attention to the use of appropriate language, understandable by all)	The project and its stakes, as well as the need it addresses, are co-constructed with the stakeholders/ public and clearly presented

Section 2 – "Step-by-step" participation

"Step-by-step" participation provides a formalised framework for the participation project, including monitoring and evaluation. Project formalisation invites the actors to clarify the subject of participation, the expected outcome of the participation, who will be involved and the expected level of participation. Project monitoring allows each participatory event to be specified: date, place, mobilised means (number of contributors, budget, and allocated time), number of participants, their age, gender, type of actor, geographical origin, as well as the means of communication put in place. Lastly, an analysis of the discussions takes into account the subjects addressed, by whom they were addressed, in relation to which objective or project, whether a consensus was reached or not and the follow-up provided.

Visualisation of the results of participation monitoring shows:
– the total number of participants broken down according to:
– types of participatory mechanisms put in place;
– type of participant, age, gender;
– geographical origin of participants: depending on the data entered, the application maps the geographical origins of the participants;
– the mechanism schedule.

The application also provides the means to track project communication:
– the amount of communication dedicated to the process:
– types of media used;
– how long it was communicated;
– the communication schedule.

Lastly, the application can be used for reporting and accountability by sorting topics according to what was agreed or disagreed upon and/or according to the follow-up. Reporting on the decisions taken at the end of the participatory process is all the more relevant if it is carried out collectively, by including all project stakeholders (as is the case for self-assessment and step-by-step project formalisation). By definition, all project stakeholders should receive the final report.

All these Participation compass features contribute to the development of a culture of participation and aim to improve regional practices.

More information and access to the participation compass: boussole-participation@cerema.fr

Chapter 11

Participating is also learning!

*Laura Seguin, Patrice Garin, Sabine Girard,
Sarah Loudin and Emeline Hassenforder*

Learning is an important effect of participatory processes. As a participant, facilitator or commissioning authority, anyone who partakes in the process acquires some form of new knowledge, know-how or skill through the social interaction that takes place. Yet, in practical terms, learning is rarely assessed. So, what exactly is involved and how should it be done? In this chapter, the authors propose a formalised framework based on their own research experience on participation in the field of water management.

▶▶ A few elements for understanding

Beyond the effects on decision-making, participatory processes can have effects on the actors who take part in them: citizens, professionals, elected officials, community organisation members, researchers, etc. As bearers of knowledge, interests and different representations of the issue under discussion, actors come together to put these factors forward, to guide and even transform the discussions and decisions of the issue at hand. In return, they too are transformed. As a place of social interaction, participatory mechanisms are spaces for learning, including the (trans)formation of understanding, individual knowledge, know-how, skills, and at times social representations and behaviour.

Why show interest in these learning processes?

Firstly, to be able to establish causal links between what happened during the participatory experience and decision-making. Establishing a new collective management rule, for example, will have required several acquisitions: new knowledge or ways of perceiving an issue at the diagnostic stage, knowledge of other participants, establishment of rules regulating discussion between them, debating skills and the ability to establish a collective opinion, to work towards a common interest, etc. Identifying these links means showing the added value of participation with regard to decisions taken by a panel of representatives who have in this manner acquired political skills and the ability to inform themselves, debate and decide. It also means identifying the key points to be worked on with the audiences not directly involved in the process, so that they too understand the decision taken.

Acquired knowledge and know-how can then be reinvested elsewhere, in other situations, and feed other forms of democracy. For example, citizens who have acquired new skills during a participatory process may then become involved in a cause, or in local politics, strengthened with what they have learned (Talpin, 2011; Seguin, 2020). A social extension of learning may therefore occur, both over time and within different social groups: the individual, the group of participants, the social groups to which they belong, society at large or local institutions, etc. (Webler *et al.*, 1995; Reed *et al.*, 2010)

Learning: what is it about and how can it be assessed?

Participation is a social activity that tests the individuals and groups that take part in it. It can be seen as a series of events that constitute learning situations (i.e. workshop discussions, field visits, surveys, exercises to explore possible futures or experiments, time for reflection, etc.). The questions summarised in figure 11.1 can be used throughout the process to grasp the effects of this learning.

Who learns?

By considering the learners in a participatory experience through a broad lens, we may consider both the mandators and designers of the mechanisms (elected representatives, institutional actors, researchers, facilitators, etc.) as well as the participants mobilised (socio-professional actors, community organisations, citizens, residents, locals, etc.), and even audiences not directly involved in the process, but who may be affected indirectly via social networks. Let us bear in mind that the boundaries between these categories are fuzzy: an elected official, institutional, socio-professional or community actor may be both a mandating authority and a participant. Citizens or local residents, on the contrary, rarely initiate or design participatory approaches. And while participation is still too often thought of as a one-way "educational" tool for citizens, this broad view demonstrates that it can also be a rich source of information for public authorities and stakeholders.

Individual or collective learning?

A distinction can be made between learning at the individual level and learning that takes place within a group that has been or is being formed during the participatory experience. Work within a group particularly leads to transformations in the ways in which we learn together, i.e. confronting each other's views in order to enrich each other's skills and develop a common capacity for action. Thus, the collective development of expertise or the gradual establishment of discussion rules allowing everyone to participate are examples of collective learning. Moreover, the direct participants are not the only ones who learn; forms of dissemination through social networks may appear, for example via an organisation or club to which they belong, or through the organisation in which they work.

What is learned?

Learning differs depending on its nature. Cognitive learning refers to knowledge; it can be expert, professional or practical. Political learning refers to the acquisition of skills, know-how or aptitudes that encourage involvement in collective action i.e. taking the floor, listening, debating, generalising, leading a discussion, managing conflictual negotiations, formulating an opinion, etc. Organisational learning refers to the construction of new forms of organisation and/or exchanges between actors.

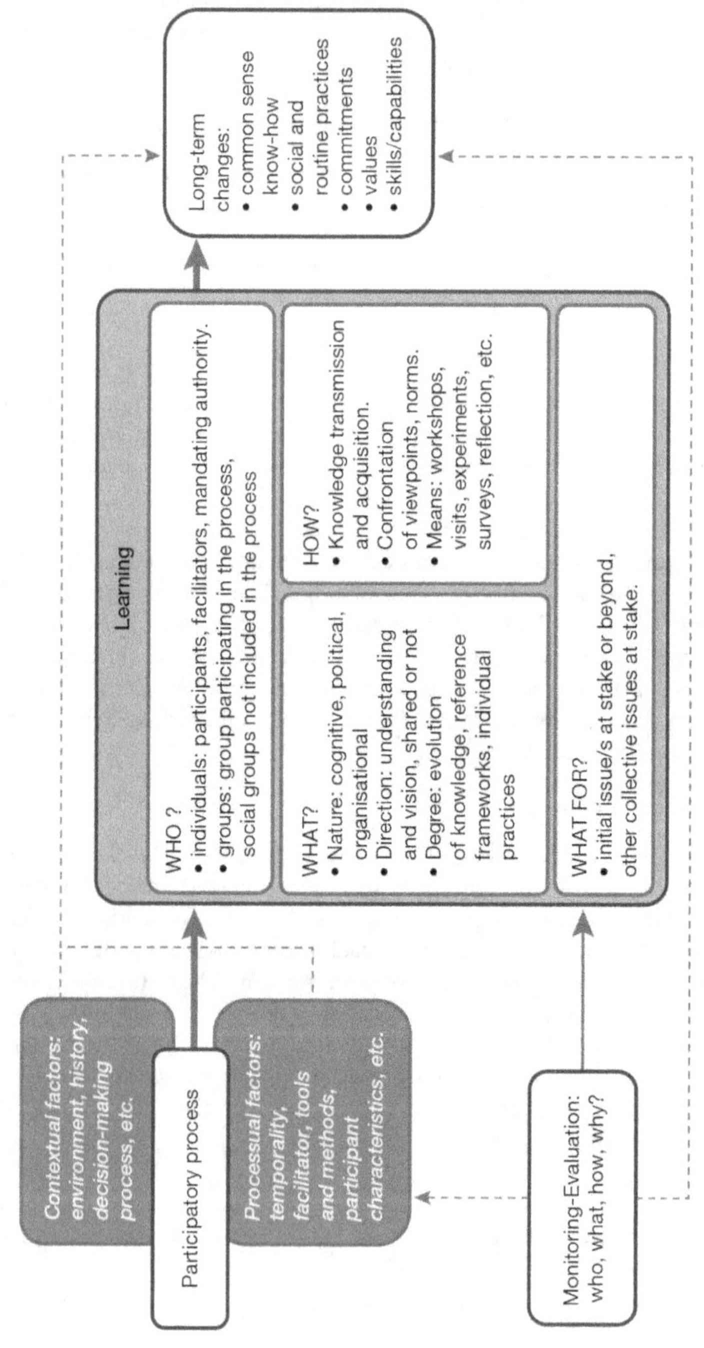

Figure 11.1. Conceptual framework for the assessment of learning

The direction learning takes within a group can also be of interest: does it allow for the co-construction of shared understanding or, on the contrary, does it reinforce divergent views?

How does learning take place?

Whether individual or collective, learning is part of social interaction. However, participants can play a more or less active role. There are different types of learning events or situations such as the transfer of knowledge (e.g. when an expert informs participants), debate and negotiation (e.g. when participants exchange with each other), inter-comprehension (e.g. when each participant explains what they see and understand during visits), investigation or experimentation (e.g. when participants are asked to identify a problem and to find the solution themselves) or reflective feedback (e.g. when a group is led to reflect on what it has learned and what it lacks for the future). These situations do not produce the same types of knowledge. However, the importance of such learning is elicited well in the words of philosopher and teacher John Dewey: "learning by doing".

The methodology used to assess learning can be broken down into four main elements:
– *When to assess?* The temporal dynamics of all these effects require several moments for observation: before the participatory process (ex-ante), during it (in itinere), just after it (ex-post), or even a long time afterwards (a posteriori).
– *How to assess?* Among a wide range of tools, two main categories can be mentioned: external assessment (interviews, cognitive mapping, participant observation) and self-assessment involving reflexive feedback from participants (through self-administered questionnaires, workshop debriefings, role-playing using serious games, or viewing key moments that have been filmed).
– *Who assesses?* As with the whole process (see chapter 10), mandating authorities and participants can contribute to the assessment of transformations; firstly, by identifying topics on which they need to deepen their knowledge or develop their skills; and secondly through self-assessment of the transformations or through peer evaluation.
– *Why assess?* The aim is to identify any shortcomings in the system for the participants as well as to identify the key points of the group's dynamics (a shift in views, reaffirmed opposition to certain points of view, a lack of knowledge or skills for some). These elements can help to plan a complementary action aimed at audiences who did not participate. From a scientific standpoint, evaluation sheds light on the mechanisms and comparative effects of different approaches.

What connections can be made between participatory approaches, learning and long-term change?

The objective of a participatory process is often transformative: it aims to influence behaviour or implement actions in response to a given regional issue and to which a group of stakeholders is trying to respond. It is also about strengthening the capacity of actors to adapt to the challenges that will arise in the longer term and on different scales.

However, translating this learning from a small group to large-scale societal transformation at the local level involves long and complex processes:
– the knowledge acquired by the beneficiaries of the scheme will clash with the common sense knowledge of the social groups to which they belong. Accompanying actions can help the knowledge acquired by a few to trickle down to the masses (e.g. environmental education, etc.);

– the acquisition of knowledge and a new shared social norm on behaviour deemed "virtuous" do not necessarily translate into changes in daily practices. Including them into routine practices is a complex individual and social process (examples: sorting waste, reducing water consumption, etc.);

– extension of engagement in collective action is also multi-dependent. If a participant becomes involved in a local organisation or in the town's administration, this is undoubtedly linked to the acquisition of new political knowledge and know-how, and probably also to the fact that this learning is coupled with an effect of social recognition and promotion. This effect is itself to be crossed with the socialisation effect that the experience has allowed e.g. the meeting of community actors or elected officials, the possible bonds of trust or even friendship that have been established, etc.

▸▸ Feedback

Table 11.1 and boxes 11.1 to 11.5 present five assessment take-aways from learning that ensued from participatory processes. In order to compare them, we have used the theoretical framework elements set forth in the previous section. Consequently, these examples illustrate the diversity of learning audiences, the types of learning, their modalities, as well as the diversity of the methods used to assess this learning.

Table 11.1. Comparative presentation of case studies

Participatory process	Example 1: Citizen conference on water within the scope of a water management plan (SAGE[1]) in Charente	Example 2: Participatory research: Brie'eau project	Example 3: Preparation phase for revision of the Drôme water management plan (SAGE[1]) with the citizens (SPARE[2])	Example 4: Testing water scarcity learning devices on the Drôme and Cèze rivers	Example 5: Ex-ante ex-post assessment of capabilities using a role-playing game (CappWAG
Who has learned?	Randomly selected citizens, elected officials, water managers, CBOs[3]	Farmers, agricultural cooperatives, elected officials, water managers, CBOs, researchers	Citizens (voluntary participation), water managers (river syndicate agents; local water commission members); researchers	Two panels of about ten elected representatives and ten citizens	Approx. 40 students at Master 1 level
What was learned	Cognitive knowledge on water, the catchment area, use conflicts and participation processes. Participatory know-how (workshop)	Cognitive knowledge on the agricultural system of the region, pollution. Know-how (participatory and organisational).	Cognitive knowledge on water, its uses, its management and on participation. Relational knowledge (who the others are and how to do things together); Participatory and organisational know-how	Essentially cognitive knowledge on water, drinking water, its stakes, its management, pricing, water scarcity and water scarcity management, the territory.	Political knowledge (taking the floor, making a diagnostic together) and organisational knowledge (creating and implementing new management rules).
How was it learned?	Collective survey, role-playing on a watershed, hearing from experts, working in small groups, writing a collective opinion.	Small group workshops, field visits, participatory simulation workshop (role-play).	Series of participatory workshops: focus group, expert hearings, modelling, role-play, survey, etc.	Focus groups, dialogue with experts; serious game on pricing, development of communication materials.	3-month course on participation combining theory and case study group work.
How was it assessed?	Interviews, observation, video feedback.	Interviews, observation, collective debriefing at the end of workshops.	Interviews, observation, self-assessment, group debriefings.	Interviews (keywords, drawings and questions), self-assessment.	CappWAG assessment tool: role-play, questionnaire, focus groups and interviews.

Participatory process	Example 1: Citizen conference on water within the scope of a water management plan (SAGE[1]) in Charente	Example 2: Participatory research: Brie'eau project	Example 3: Preparation phase for revision of the Drôme water management plan (SAGE[1]) with the citizens (SPARE[2])	Example 4: Testing water scarcity learning devices on the Drôme and Cèze rivers	Example 5: Ex-ante ex-post assessment of capabilities using a role-playing game (CappWAG
When was it assessed?	Before, during, just after and two years after the participatory process.	Before, during and just after the participatory process.	Before, during and just after the participatory process.	Before, during and just after the participatory process.	Before and just after the course on participation.
Who evaluated?	An external researcher.	The steering group of the research project (researchers and their partners) by involving the participants.	A monitoring and evaluation group composed of researchers, project leaders and citizens.	The researchers.	The researchers.
Why was it evaluated?	Knowledge of learning processes (research).	Knowledge and adjustment of the process along the way.	Awareness raising and empowerment of actors in terms of participation (and its assessment) Knowledge on learning processes.	Knowledge on learning (effect of the interaction devices tested).	Knowledge of learning/ capability acquisition processes.

[1]SAGE: schéma d'aménagement et de gestion des eaux.
[2]SPARE: European project "Strategic Planning for Alpine River Ecosystems 2015-2018".
[3]Community-Based Organisation

Box 11.1. Learning from a citizen conference in the scope of the SAGE Charente

In 2011, the Regional public establishment (EPTB*) for the Charente river in France—the structure responsible for the then emerging local water commission—joined forces with a regional organisation (Training and research institute for environmental education—Ifrée**) to implement a process to involve residents from the catchment area in the local water policy.

The approach included three weekends that first provided citizens with information (on water management, the notion of the water cycle, the watershed, as well as on the conflicts of use in their region), then included them in critical investigation during meetings with experts, stakeholders or users of water, and finally provided a space for deliberation and construction of proposals in a workshop aimed at drafting a collective opinion.

The citizens' diagnostic and their proposals for water use conflicts set the tone for the work to be undertaken within the local water commission. This experience was monitored in order to identify the learning processes at work in the various actors who took part: elected officials and EPTB agents, Ifrée facilitators and citizens. Interviews were conducted before, just after, and two years after the process in order to identify long-term learning. These interviews were complemented by observation notes taken during the steering committee meetings, each weekend workshop, and various discussion times that led to drafting of the opinion. In a rather original way, video was used as a methodological tool in order to collect the participants' feedback of their experience in the experiment, and to identify what they had learned.

The results show cognitive and political learning. Firstly, in the water managers who, together with Ifrée facilitators, gradually became acculturated to a different way of conceiving public participation. Secondly, in the participants, who, in addition to having built up group expertise on the issue, acquired keys to political interpretation that they did not have before, feeding curiosity that in turn transformed their habits on obtaining information, for example, and even politicised certain individuals. These effects, which are still visible two years after the experience, are sometimes reflected in continued involvement (in CBOs, activism, local politics; Seguin, 2020).

* Établissement public territorial du bassin.
** Institut de formation et de recherche en éducation à l'environnement.

Box 11.2. Regional dialogue on the issue of diffuse agricultural pollution (nitrates, phytosanitary products) in the Brie region

The Brie'eau research project aimed to experiment with a participatory approach to facilitate dialogue on diffuse pollution of agricultural origin in the Brie region of France (Seine-et-Marne department). Farmers and stakeholders from the agricultural sector, local elected officials, drinking water stakeholders and local user CBOs were encouraged to co-construct a more resilient region by using two levers for action: changes in agricultural practices and landscaping that acts as a buffer zone by intercepting part of the pollutants between agricultural plots and the surrounding environment.

Box 11.2. (next)

A card game adapted from a pre-existing game, Mete'eau (Barataud *et al.*, 2015), was used to highlight perceptions and values, which were then to be discussed, of each actor from the area and concerned by its issues. This phase was followed by field visits and exchanges with scientists, which were conducive to knowledge sharing. A simulation tool was then used to build a common vision of the region and to imagine evolutionary agronomic scenarios. Finally, a role-playing game built with the help of the Lisode consultancy firm provided a virtual space for discussion and negotiation around individual and collective actions (figure 11.2).

The entire process was observed, and interviews with the project initiators (scientists and their partners) and the participants were conducted before and just after. The participants testified to the acquisition of knowledge on the issue of water quality, on buffer zones and their multiple functions, and on the agricultural system of their region. Moreover, this experiment has contributed to the creation of a community of concerned stakeholders, who know each other better, who are able to hear each other's different visions and who are ready to continue the reflection together. This first step was essential to the sensitive and contentious subject of diffuse agricultural pollution.

Even if it is still too early to talk about real organisational learning, several signs point to a shift in the way local collective action on water and agricultural issues is thought out (Seguin *et al.*, 2021).

Figure 11.2. "Res'eaulution Diffuse" role-play (Brie'eau project)

Box 11.3. Citizen participation in the preparation of the Drôme SAGE revision

Water management plans (SAGE), the main planning tool for water management at the local level, are drawn up by local water commissions, which include representatives of State services, elected representatives and users. But what about the citizens and local populations of the concerned catchment area?

Between 2016 and 2018, as part of a European research-action project (SPARE Project, Interreg Alpine Arc), the "Syndicat Mixte de la Rivière Drôme" union decided to collect the opinions and proposals for action of citizens on the river and its management in order to feed this into the revision of the SAGE.

The originality of the approach was to involve participants in the design, implementation and monitoring-evaluation of the participatory process itself. Thus, the expected learning was as much about the subject of water and its management as about the subject of citizen participation itself (how to do it, for what purpose?).

The learning was assessed through participant observation, semi-structured interviews and self-assessment questionnaires.

The results show cognitive learning by the citizens, in particular on water, its uses, the stakes and the organisation of water management, as well as on the room for citizen action. Some of the proposed actions thus concern access to information and the possibilities of contributing more actively to local water governance. This learning is also relational and organisational; for example, it has led to the integration of citizen as participants in the local water commission.

Learning, notably organisational learning, also took place among the agents and elected representatives from the river union: the latter modified their communication policy, internal working methods and facilitation of the Drôme SAGE (Ferrand *et al.*, 2018).

Box 11.4. Learning about water scarcity in the Drôme and Cèze regions

Adapting to climate change requires a change in consumption practices, especially for drinking water. These behaviours depend on the representations that each person makes of their practices and their effects on the environment. They are qualified as common sense knowledge, which is transformed in places of social interaction and via the media where perceptions, attitudes, experiences and opinions are encountered. Participatory mechanisms can be considered as times when points of view are confronted. They are said to have the capacity to promote the dissemination of new social representations, but in reality, how true is this?

Exactly this is what was tested in a research project financed by the Rhône-Mediterranean-Corsica water agency. The research team traced the evolution of social representations of water, drinking water, its scarcity as well as the way it is managed, following each of four interactive sessions that took place:
- focus group,
- dialogues with experts,
- exploration of social dilemmas on water pricing in the course of a serious game,
- collective elaboration of communication materials for the general public.

Box 11.4. (next)

The project mobilised four groups of elected representatives and citizens from the Diois greater municipality (Drôme department) and the Cèze-Cévennes greater municipality (Gard department), located in water distribution zones.

Four methods were used to analyse these representations and their evolution: i) the associative method based on the statement and classification of word-images by each person in reaction to hearing a word-inducer (water, drinking water, scarcity); ii) mind maps on the issue under discussion (drawing where the water from my tap comes from or where it goes next); iii) semi-structured interviews and iv) individual and group self-assessments.

The results show that the serious game on water tariffs and the collective development of communication materials were the two most fruitful mechanisms for social learning. Climate change and its concrete consequences had the greatest impact on people. However, while knowledge evolved, there has been little concrete change in consumption practices or in the way services are managed. Identifying the conditions required to translate this new knowledge into new practices would require extending the study into looking at the motivations for acquiring new equipment and making new behaviours routine.

It could draw on recent developments in the theory of practices applied to the sociological study of consumption (Garin *et al.*, 2022).

Box 11.5. Ex-ante ex-post assessment of capabilities using a role-playing game (CappWAG)

For a participatory process to be transformative and effective, participants need to have a number of participatory capabilities, i.e. capacities to participate (Frediani, 2015). These correspond to potential capacities for action that allow them to take part in the participatory process in possession of all the necessary means to make their voice heard and to have influence. These skills to be acquired in order to make an informed contribution to the decision-making process are similar to political or organisational learning, among other things.

In order to assess the existence, strengthening or weakening of these capabilities, the CappWAG assessment tool was developed (figure 11.3). It is based on an eponymous role-play (divided into an ex-ante and an ex-post version), a questionnaire and a collective debriefing. The tool was thus implemented in 2017-2018 to evaluate the impacts of a three-month course on integrated water resource management with five groups of first-year Master's students.

The results showed that learning of the three assessed skills (speaking in front of a group, making a collective diagnostic, and creating and implementing management rules) was very diverse. After the three-month course, these had not always increased in the groups (expected impact) and sometimes even decreased! The course, but also socialisation of students elsewhere outside of the university, were cited as the main factors influencing the individual and collective abilities of the students to work together.

This case study thus allowed for a better understanding of how political or organisational learning is formed and evolves over time (Loudin, 2019).

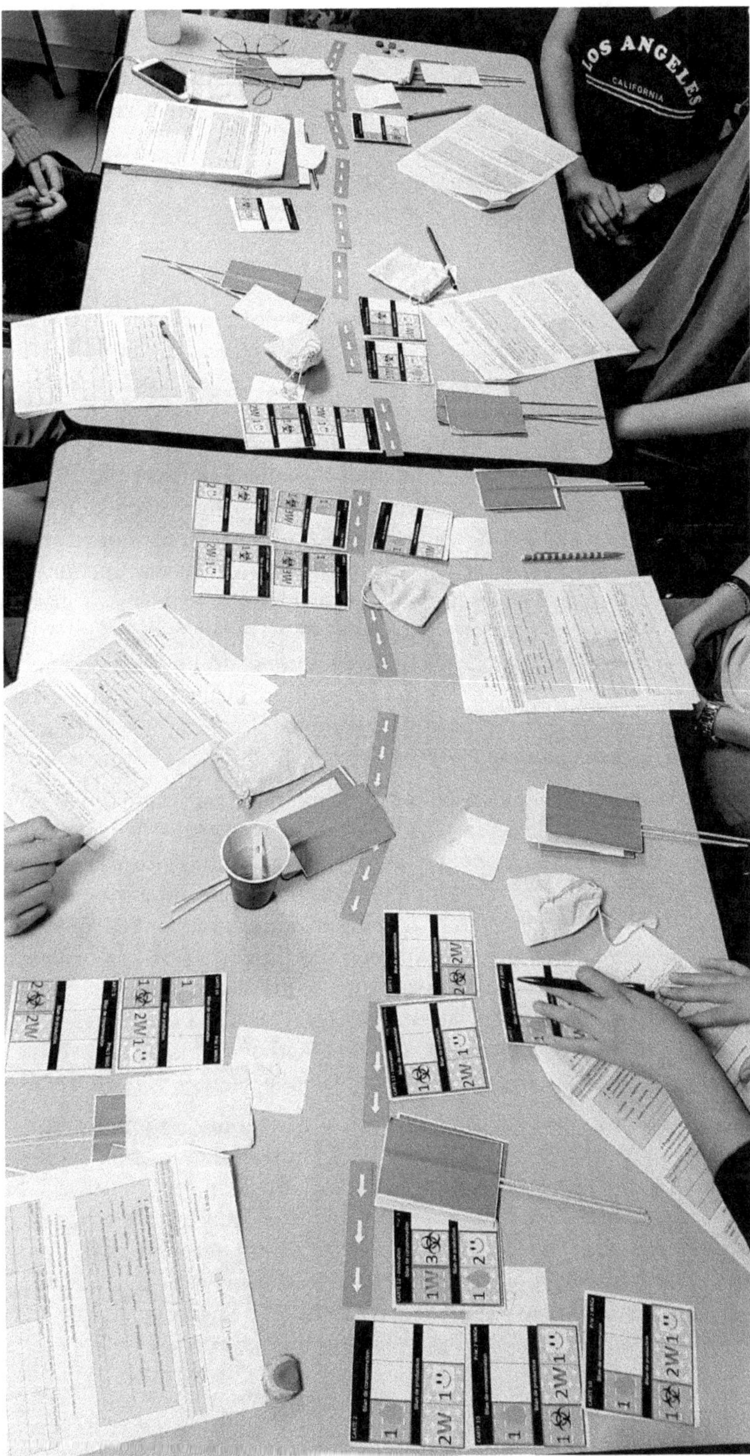

Figure 11.3. CappWAG, a tool for assessing learning and skills

▸▸ Conclusion

A participatory process is always rich in learning and constitutes just one more reason to fully commit to it. Sometimes, learning may even be the main objective of the project initiators.

Individual and collective investigation constitute a formative experience, both for the participants and the mandating parties, as well as for the facilitators who can contribute to the co-design of the process, its implementation and its evaluation.

However, learning depends on several factors. The first factor is temporal: mobilising people over a long period (as in the example of the Drôme SAGE: two years) can allow for more intense transformations to take place than an ad hoc and time-limited approach (as in the case of the citizen conference: three weekends and two days of feedback). Moreover, the initial knowledge and skills of the participants also have consequences on the types and paths of learning. In the course of our experiments, we have observed, for example, that a participant, who is already involved in community-based organising and already familiar with the functioning of public action and the art of negotiation, will more readily be at ease with the functioning of water policies and will feel comfortable in a workshop format that encourages debate. This is why the sociological characteristics of the mobilised audience (initial training, profession, commitments, previous participation experience, etc.) are an important factor to take into account when seeking to promote and/or evaluate learning. In this respect, we note the importance of the role and skills of facilitators and coaches, as well as the tools they use, in creating the most favourable situations for the transformation of knowledge, depending on the diversity and unequal abilities of the participants.

In the examples presented, participation professionals were sometimes called upon for their facilitation skills (Ifrée, Lisode). They contribute to the pedagogical aspect of materials, presentations and visits, facilitate exchanges and ensure the participation of everyone, in particular by taking into account the pre-existing power relationships between participants. Finally, these transformations have long-term effects; they deserve to be observed, not only during and after the process, but also long after the participatory process has ended. The five examples presented show that the assessment of learning can be carried out by a variety of actors (researchers, project steering group, the participants themselves) and using different methods. This can only be achieved if the necessary resources are anticipated right from the beginning at the design stage.

The question of learning is an integral part of the monitoring and evaluation of a participatory approach. Some of the methods proposed in chapter 10 are adapted to this objective. Let us bear in mind that the gains of an experience will be all the stronger if they are identified by the actors themselves. It is therefore important, even if an external person is called in, to share this assessment with all the participants and to allow them to reflect on what they have learned as individuals and as a group. Learning to learn, learning to pay attention to what is being learned, why and how we learn, is an important step to learning, whatever the subject may be.

⤻ References

Barataud F., Arrighi A., Durpoix A, 2015. Mettre cartes sur table et parler de son territoire de l'eau : un (en)jeu pour les acteurs ?, *VertigO - La revue électronique en sciences de l'environnement*, 15, 3. https://id.erudit.org/iderudit/1035877ar

Ferrand N., Girard S., Hassenforder E., 2018. *Implementation and results of monitoring and evaluation methods. Participatory processes for strategic planning of five alpine rivers.* Spare Project, final Report, DT132, 106 p. https://www.alpine-space.eu/projects/spare/downloads/last-publications-from-boku/spare_implementation-and-results-of-monitoring-and-evaluation-methods.pdf

Frediani A. A, 2015. "Participatory Capabilities" in Development Practice. *In:* Frediani A. A., Hansen J., (éd.), *The Capability Approach in Development Planning and Urban Design.* London. Development Planning Unit, The Bartlett, University College, 121-132 (Coll. DPU Working Papers Special Issue).

Garin P., Girard S., Rivière-Honegger A., Degache A., Gouton C., Pellen M., 2022. Running out of tap water one day? Territorial distancing from the effects of climate change. *Géocarrefour*, 96 (96/1). https://doi.org/10.4000/geocarrefour.18281

Loudin S., 2019. *Can We Use a Social Experiment to Assess the Impact of Participatory Processes for Water Management? Studying a Generic Method Tackling the Evaluation of Capabilities*, Thèse de doctorat, spécialité Sciences de l'eau, AgroParisTech, Paris, 359 p. https://www.theses.fr/s211236.

Reed M. S., Evely A. C., Cundill G., Fazey I., Glass J., Laing A., *et al.*, 2010. What is social learning?, *Ecology and society*, 15 (4). http://www.ecologyandsociety.org/vol15/iss4/resp1/

Seguin L., 2020. *Apprentissages de la citoyenneté : expériences démocratiques et environnement.* Paris, Éditions de la Maison des sciences de l'homme, 360 p. (Coll. Le (bien) commun)

Seguin L., Barataud F., Guichard L., Bonifazi M., Souchère V., Bouarfa S., *et al.*, 2021. La participation comme objet intermédiaire d'apprentissages : leçons d'une démarche participative sur les pollutions diffuses agricoles. *Natures, Sciences, Sociétés*, 3 (29), 299-311. https://doi.org/10.1051/nss/2021062

Talpin J., 2011. *Schools of democracy. How ordinary citizens (sometimes) become competent in participatory budgeting institutions*, Colchester, ECPR Press

Webler T., Kastenhol H., Renn O., 1995. Public Participation in Impact Assessment: A Social Learning Perspective. *Environmental Impact Assessment Review*, 15 (5), 443-463. https://doi.org/10.1016/0195-9255(95)00043-E

Part 3

Participatory Modelling to Support Decision and Change

Chapter 12

Designing and using role-playing games for water and land management

Géraldine Abrami and Nicolas Becu

This chapter deals with the design and use of role-playing games as methods for implementing participatory approaches to water management and more broadly to land management. It addresses various methodological points about this approach in the form of questions and answers, and then presents the kit for designing the participatory role-playing game Wat-A-Game. As a comparison, a glimpse into LittoSIM, a game platform dedicated to coastal flood risk prevention in France, is also provided.

A role-playing game is made up of material elements (boards, cards, tokens, computer interfaces, etc.) associated with a system of rules that define participants' interactions with the game, as well as game dynamics. In the context described in this chapter on how the game is used, the material elements and the rules are the translation into a gaming form of a model of socio-environmental dynamics and interactions at work in a territory.

It is useful to distinguish between:
– the role-playing tool (the artefact and the way it is used in a workshop);
– the process in which it is used or the deployment of the game in a territory (the broader participatory process and the partnership that drives it);
– the social and territorial context (institutional or citizen processes, arenas of action, etc.).

▸▸ Historical background

The use of games to support land management processes dates back to the 1960s and 1970s, when numerous games were developed based on systems thinking and complex systems to represent systems of interactions between resources, uses and modes of regulation (Duke, 2011; Klabbers, 2009). Two major scientific communities contributed to the development of this type of tool. In the field of computer sciences, a trend in modelling complex systems in ecology and economics was initiated through the development of dynamic systems (see chapter 2). In the field of communications and management, the simulation and gaming community focused on the emotional, sensitive and interpersonal dimension of the collective involvement of groups of actors in interactive simulation experiences (Duke, 1974; Klabbers, 2009). This research has stabilised methods and principles of play that are used today in training, teaching and coaching sessions in various fields of application (medical operations, project and human resource management, military strategy, risk management, foreign language learning, etc.). They have also led to crossovers with other gaming-inspired approaches, such as the current trend towards the "gamification" of online service platforms.

Several decades after these initial experiences, the "Companion Modelling" (ComMod) movement emerged in France in the early 1990s. This trend developed a new approach to the use of models and games in the field of renewable resource management. The originality of this approach is that it proposes a complete participatory approach, from the initial analysis of the actors' network to the evaluation of the effects of the participatory process. With a participatory aim, it applies modelling and simulation through games as the framework for reflection and dialogue in order to explore and discuss possible futures in a group setting (Mathevet and Bousquet, 2014; Ostrom, 1990). Starting in 2008 and looking to bring the ComMod approach to a large audience in which the actors could independently apply the approach to water management, reflections undertaken by the G-EAU joint research unit "Water Matters", gave rise to the Wat-A-Game platform and the CoOPLAGE approach (see chapter 2).

> **Box 12.1. Particularities/Features of role-playing games compared to other participatory tools**
>
> Role-playing is:
> - ergonomic: adapts to different types of actors;
> - playful: creates distance thus facilitating exchange and reducing tensions;
> - experiential: mobilises emotional intelligence, tacit knowledge and allows for usual analysis frameworks to be surpassed.
>
> These specificities produce effects that can be classified into three types:
> - create a space for exchange and interaction between participants;
> - generate learning (see chapter 11);
> - cause changes in the perception of the represented territory.
>
> Role-playing within the scope of a Companion Modelling posture have two additional specificities:
> - they are physical or computer-based models that simulate the responses of the environment to the decisions and interactions of the participants;
> - their design and use are part of participatory approaches.

▸▸ Wat-A-Game: a kit for designing role-playing games for water management

Wat-A-Game (WAG) is a family of methods and tools that allow, with minimal training, groups of actors of all levels to build non-computerised role-playing games on water and land management. WAG provides material building blocks associated with standard usage rules, a collective design protocol, and a library of prototypes and games. WAG is part of the CoOPLAGE suite, and is particularly complementary with the participatory strategic planning tool CoOPLAN (see chapter 2). WAG can thus be used to place actors in situ before a strategic planning exercise; it can then be adapted to explore strategies resulting from this exercise.

Access to WAG methods and tools is associated with a user agreement and online registration on the CoOPLAGE website. More information at http://cooplage.org/tools/wag-and-its-family

The bricks and principles of Wat-A-Game

The water resource, which flows through networks where it is subject to abstraction and discharge, lends itself particularly well to representation in the form of beads that circulate in containers from which players can draw and which can collect discharge (figure 12.1). This is how beads of different colour became one of the basic building blocks of WAG, in which they represent water or other types of resources. Other blocks include activity cards representing uses characterised by the production and consumption of different beads, plots of land on which activities are carried out and which induce modes of access to resources, notably via their connection to a simplified hydrographic network, and finally roles that specify access rights on the plots of land, possible activities, as well as specific objectives and constraints.

A WAG game round consists of a first phase in which players take individual or collective decisions on their activities, infrastructures and management rules, followed by a resolution and reporting phase. The resolution consists in manually moving the beads according to rules that represent the dynamics of the resources, as well as the consumption and production parameters set out in the activity cards. The assessment or reporting is done by filling in a monitoring table.

Many phenomena can be represented using this simple base, for instance in the biophysical field, phenomena may include: run-off, infiltration and flooding, pollution with dilution, sediment transport or biodiversity evolution, or in the social or economic field, phenomena may include: corruption, advocacy, demography, or multiple land tenure.

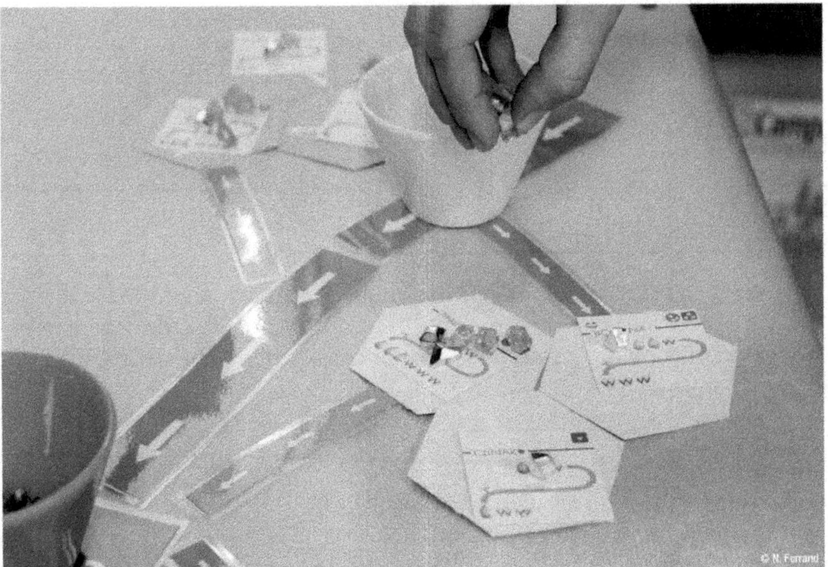

Figure 12.1. Wat-A-Game blocks in action (here the INI-WAG game): a player takes water for his activities from the cup that runs along the river

The Wat-A-Game design protocol

Wat-A-Game provides a stabilised collective game design protocol for two half-day sessions, which can be used to produce a co-constructed prototype that incorporates the issues and perspectives brought to the table by the group's participants. For a shorter time period, two to three hours suffice to discover the basics of socio-environmental modelling and to obtain an initial rough prototype using a "design-by-playing" approach. These two protocols are summarised in table 12.1.

Table 12.1. The six steps to collectively designing a game protocol (CreaWAG).

CreaWAG protocol phases	In full mode	In design-by-playing mode
0. Modelling group engagement	Can be done through immersion in an introductory game	Idem
1. Framing	Specification of use, issues and scales and indicators associated with the issues	Specification of an issue
2. Conceptual modelling	Spatial structuring of the territory, inventory and selection of actors and resources, analytical mapping of links between actors, issues and resources	Rapid spatial structuring, identification of two or three resources and one role per participant
3. Prototyping and pre-testing	Taking on a role, specifying its objectives and activities, designing the board and simulating a game round	Identification of three to five activities per role and on-the-fly calibration during the pre-test
4. Finalisation	Review and specification of the board, roles and activities, resource dynamics, scenarios and events, and the flow of a game round	Not finalised
5. Calibration	Calibration of activities and scenarios with abstract and simplified values and relative orders of magnitude	No calibration
6. External test	Testing the prototype with an external group	No external test

Since Wat-A-Game's debut in 2008, the CreaWAG protocol and its design-by-playing variant have been more or less fully implemented over one hundred and fifty times in various international contexts and often in conjunction with other tools from the CoOPLAGE suite (Figure 12.2).

▸▸ Ready-to-use games from the Wat-A-Game family

Two introductory games are currently available:
– IniWAG is an abstract game on quantitative and qualitative resource management;
– MyRiverKit is a generic game set in a European context that aims to raise awareness of integrated management concepts such as ecosystem services. It was designed to be played independently, without a facilitator.

Figure 12.2. CreaWAG design workshop

These two games provide an abstract or generic framework for experimenting with typical collective water management scenarios and are well adapted to launching a design-by-playing workshop. Several hundred copies have been produced and are available in physical form.

Other ready-to-use games address more specific cases or issues, including the educational boxes "L'*Eau en Jeu*" and "L'*Eau en Têt*" (see chapter 14) and the WasteWAG sub-Saharan sanitation planning kit (see chapter 15).

Finally, most items produced, whatever their level of completion, are listed in an online database called the "WAG-lib" (https://bit.ly/thewaglib).

▸▸ An emblematic case study: Mpan'Game and the Ugandan process

"Mpan'game" (from the European project AFROMAISON, 2012-2015; see also chapter 17) is an apparatus addressing water quality deterioration and its impacts on ecosystems in the Mpanga river basin in Uganda. It was developed by a small group of stakeholders gathered around the Mountains of the Moon Community University in Fort Portal, animated by the G-EAU team. After two days of training, facilitators from a local non-governmental organisation were able to deploy it in conjunction with the participatory planning tool CoOPLAN and the participatory monitoring and evaluation tool ENCORE-ME (see chapter 18) to thirty-five communities around the basin. Representatives of these communities then participated in a regional multi-level strategic planning workshop. The tool has continued to live on since then and is still used in local projects. More information in Hassenforder *et al.* (2015).

Figure 12.3. Mpan'Game session in a community in the Mpanga river basin (Uganda)

▸▸ Questions and good practices around the design and use of role-playing games

As the main elements of the WAG kit have been presented, it is now time to address any questions that an actor in the field may have and who would be interested in using a role-playing game to facilitate a participatory process in their territory; a few good practices that we feel are essential are also presented here. Box 12.2 develops how a role-playing tool can contribute to a participatory process, in terms of understanding the workings of a territory, interaction between and knowledge of the actors, strengthening a collective, or exploring scenarios and alternatives.

This section borrows heavily from Becu (2020) and Barreteau *et al.* (2021).

Turnkey or ad hoc game?

There are many turnkey games (all set, ready to use games) that correspond to different issues; however, it may be more appropriate to specifically design a game for the situation and issues of one's territory.

Turnkey games are attractive, simple and visible. Thus, allow for workshops to be rapidly implemented and if well chosen, participants can be led to address issues relevant to their territory. Well suited to awareness-raising campaigns, they can also be used as an introduction to other activities.

Box 12.2. What kind of needs can games meet?

Thanks to their instructional qualities, role-playing games are now recognised and widely used in educational applications (see chapter 14). They can also be used to meet management objectives that go well beyond raising awareness:
– to bring actors to think about the complexity of a territory (system-thinking) by experimenting, for example, with interdependencies, the effects of competition on limited resources or the need for coordination;
– to make players aware of the diversity of views, constraints and interests through interaction with other players;
– to enable a group to acquire communication and collaboration skills (argumentation, deliberation, communication at different levels, etc.) in order to increase their capacity to work together and participate in management arenas;
– to generate and experiment with alternatives. The game is a "safe" space where actors can experiment with strategies, conflicts, forms of collective action, information sharing or coordination, without having direct consequences in real life.

Designing an ad hoc game however provides the opportunity to legitimise and better integrate territorial issues through the use of the game. This entails a long-term approach that structures the commitment of territorial actors in an ambitious participatory process. Game construction workshops are important times for exchange and can promote positive group dynamics, thus strengthening the group.

Today there are modular games that fit somewhere in between these two, in which the basic elements and model corresponding to a field of application can be partially adapted using various game elements fitted to the territory. The LittoSIM game and Wat-A-Game, as well as the TerriStories kit (by D'Aquino, 2016) which deals with property, are examples of this type of device.

With or without computers?

There are different degrees and ways of computerising a game.

Computer-free games are flexible, robust, transparent and easily appropriated. Their design and use require little technical expertise. This is why the designers of Wat-A-Game have chosen to develop a participatory modelling and game design based solely on paper and beads.

A computer simulator can be used to carry out calculations and display results, as well as to capture the actions of players, alongside a physical board and game pieces, or without them (box 12.3). Computers can thus simplify the logistics, but more specifically can be used to explore more possibilities thanks to faster calculating times and multiple visualisations. The use of interfaces may hinder interaction and social learning to some extent, but this can be counterbalanced by interface ergonomics and workshop facilitation (Becu, 2022).

Lastly, there are fully computerised games that can be used remotely and asynchronously. It is however reasonable to question how much social learning takes place without direct interaction between the players.

Box 12.3. The LittoSIM game

Designed to support municipalities and inter-municipalities, LittoSIM is a game for reflection on coastal development in the face of coastal flooding risks. The tool was developed by the "Littoral Environnement & Sociétés" lab through CNRS, IRD* and several French universities researchers with financial support from the CNRS, the Fondation de France and the community of municipalities on the Isle of Oléron. Using the game, participants are brought to address various aspects of risk management:

– understanding the phenomenon of flooding and the effectiveness over time of various preventive measures that can be mobilised today;

– the implementation of inter-municipal coordination for planning and risk management;

– anticipating regulatory, budgetary and administrative constraints for the implementation of a risk prevention strategy.

A LittoSIM workshop brings together between eight and twelve people and lasts three to four hours. The game combines simulation of land planning, modelling of flooding events and a role-playing simulation of inter-municipal relations. The participants interact with the simulator using digital tablets. The game actions are sent to the flooding model which consequently calculates the extent of flooding according to the hazard's intensity. The simulated results of the flooding are then projected onto a 2m x 2m horizontal screen around which the participants gather. The players are divided into four teams, each representing a municipality in the simulated area. The teams must manage and develop their own municipality, while taking into account the actions, demands and proposals of neighbouring municipalities. These interactions and compromises between the teams take place face to face, without the use of computer interfaces.

Figure 12.4. Stakeholders test alternative flood risk prevention strategies using the computerised LittoSIM device

The tool was deployed for the first time in 2017 as part of an action to improve knowledge and risk awareness provided by the Isle of Oléron Flood Prevention and Action Programme. Four game sessions were organised for elected officials and technicians of the municipalities and a final report was produced (figure 12.4). In the end, the workshops reinforced everyone's concern and understanding of flood risk management. The participants who learned the most were the local elected officials who had little knowledge of prevention strategies before the workshops. Approximately a third of the participants changed their opinion on the different prevention strategies, with a notable shift in favour of so-called soft defence strategies. The set-up was then transposed to other contexts and territories, including Normandy and the Camargue.

* CNRS: Centre National de la Recherche Scientifique ; IRD: Institut de Recherche pour le Développement.

More information at https://littosim.hypotheses.org/

How is a role-playing session organised and run?

Organising a game session requires a team, trained in the game, that can facilitate, observe and assist the players during the session. A neutral, modular and easy-to-access space and at least two to three hours is also required, including time in which to welcome, brief, simulate and debrief. Another hour or two is needed to set up the materials and prepare the team.

The initial briefing should recall the intention of the game, specify its objectives and the framework for observing and analysing collected data, and finally it should describe and explain the elements and rules of the game which will enable the players to get started.

During simulation, it is important to maintain the pace of the game. The course and scenario may need to be adjusted along the way as well.

Debriefing is crucial, even more important than the game itself, because it allows experiences to be transformed into learning and intentions for change to be developed. This phase usually begins with a time dedicated to expressing and sharing the emotions felt during simulation, followed by a list of discussion points prepared in advance, ranging from understanding the problems encountered and their causes, to the solutions to be found. Each discussion point is moderated by the facilitator who invites the participants to go back and forth between the experience in the game and the experience in the reality of the system represented.

What do you need to know before embarking on participatory game design?

Participatory role-play design requires certain skills. Game co-construction workshops require facilitation and mediation skills, in particular to allow the group to see the model being built according to the expressed points of view, while remaining focused on the issue at hand and the required level of detail. During the design and implementation stages outside the workshop, modelling skills are required.

The time and budget required for the participatory design of a game depends on the degree of finalisation as well as the envisioned strategy to valorise and distribute it.

The framing to initial prototype stages can be completed in one to two half-day workshops using the Wat-A-Game kit protocol. A fully experimental set-up, design to implementation, can be completed in three to four workshops that can be spread over several months. For a fully finished game, which can be reused and copies manufactured, additional workshops and development time are needed. Costs linked to finalising the game with graphics or game design professionals, and manufacturing costs may also come into play.

Game design can be facilitated through collaboration with research teams who are often looking for research-intervention sites. In addition, there is a growing supply of consultants and consultancy firms specialised in designing this type of tool. An array of training courses, materials and tools are also available to managers (see section "A step further...").

Good practice in designing a role-play

Good modelling practices (iterative approach, parsimony) are all the more important to designing role-playing games as the model must not only be relevant, it must also be fun, quickly appropriable and playable over short sessions, sometimes without a computer simulator.

Using the iterative approach, a game is developed through constant back and forth between framing, conceptualisation, calibration and materialisation and through frequent use with different types of audiences (actors or stakeholders, experts, scientists, etc.) in modelling workshops or participatory simulation, as well as during more informal sharing sessions. In-house testing or testing within communities of practice should be conducted at an early stage and repetitively. Indeed, in "running" the game, we may realise that something is not properly calibrated, or that certain aspects are too complex or too simplistic.

The following principles can help you keep it simple:
– keep representation of social dynamics and constraints to a minimum and open, since these will be brought to the game by the attitude and behaviour of the participants who will make links between the game elements and their reality;
– keep the level of realism and calibration light, by representing resource sharing dilemmas and constraints faced by the actors in a stylised environment (symbolic level) without however resorting to specific details;
– keep open space for decision-making, preferably individual for learning about the system, and collective for learning about interdependencies and coordination.
– Finally, let your creativity do the talking!

A few principles for deploying an ad hoc game

This last part deals with designing the participatory process in which the game itself is designed and/or implemented. As the engineering of participation is dealt with in another chapter in this issue (see chapter 9), here we will only touch on a few specific points linked to the nature of the role-playing game tool.

Participation takes place in this type of approach during design and simulation workshops; they constitute times of learning, confrontation and exchange for the participants. These participatory workshops must be close enough together over time to maintain a collective dynamic around the game in the territory, but also far enough apart to capitalise on information from one session to the next.

The number of participants in a design or simulation workshop is generally limited to around ten or fifteen people. The choice of participants depends on the specific stages and objectives of the workshops (e.g. representative of the diversity of issues and perspectives for design workshops, technical experts for validation workshops, strategic actors in the process for simulation workshops). Materials and invitations can be adapted to the targeted audience so as to overcome, for instance, any suspicion from some actors on the playful dimension of a game environment, or any difficulties with digital media for others.

Depending on the cultural and relational context, different strategies (joint workshops, parallel workshops, successive workshops) may be adopted to involve actors from different social strata and decision levels in the process. For D'Aquino and Bah (2013), it is the simulation artefact that ensures integration between the levels, since the actors at the national level work with the version of the artefact produced in the workshops organised at the local level. For Hassenforder *et al.* (2019), the artefact is used to build the capacity of the most vulnerable in specific workshops or at specific times, so that they can then integrate into multi-level discussions.

As the game-based approach takes place in parallel with the decision-making processes underway in the territory, it is important to ensure that they complement each other. The emergence of impacts on the system can be encouraged via two important levers: the dissemination strategy to promulgate the participants' experience and the workshop productions to a wider audience; and close monitoring and evaluation to identify and take advantage of political windows of opportunity that can lead to the concrete intentions for action, or potential compromise situations that emerged during the workshops.

▶▶ Conclusion

We hope that this chapter will have given the reader a taste of what a participatory approach using role-playing can be, and what it can bring to water and land management projects.

The experiences of action-research accumulated over the last thirty years, particularly by the Companion Modelling community, have shown the interest of this type of tool in promoting the integration of knowledge and social learning within networks of actors.

Part of the current research efforts is focused on their transfer and appropriation by managers, as illustrated by the LittoSIM and Wat-A-Game set-ups. Consequently, LittoSIM is evolving into a modular platform that allows managers to adapt the elements of the flood risk management simulation to the territory of application. The collaborative design protocol of Wat-A-Game is now stabilised; the platform however continues to be the subject of research aimed at improving its appropriation and integration with the other tools of the CoOPLAGE suite (see chapter 2), and this in particular via the development of the digital CoOPILOT system.

Box 12.4. For more on this subject

The ComMod website (https://www.commod.org/en/case-studies) provides a library of case studies (including eighteen on water management), as well as methodological guides produced by Companion Modelling practitioners and which touch on the ComMod approach in its entirety, the ARDI participatory modelling methodology, and finally the observation of role-playing games.

In the Terr'Eau & co online course, two modules are dedicated to role-playing. Module 2 explains how to use the IniWAG initiation game. Module 4 allows you to follow the CreaWAG collective game design protocol. Presentation at http://cooplage.org/tools/mooc

ComMod and CoOPLAGE trainings are organised on a regular basis. See ComMod and CoOPLAGE websites.

Communities around games and participatory approaches exist.

The simulation and gaming international community meets annually at the ISAGA conference, since the first edition of the conference back in 1970. The reference journal is "Simulation and gaming", from Sage editor.

The *Jeux et Enjeux* meetings (http://jeux-enjeux.blogspot.com/; https://www.polytech-lille.fr/jeux-enjeux-2022/) have three times brought together French practitioners of these approaches. The community of practice for designers of participatory approaches in Montpellier (https://participmontpellier.wordpress.com/) offers a space for testing approaches that make extensive use of role-playing.

Finally, a 2008 report summarises role-playing and water management: Dionnet M., *Les jeux de roles: concepts clés et perspectives pour la gestion de l'eau*, LISODE. http://www. lisode. com/nos-publications.

▸▸ References

Barreteau O., Abrami G., Bonte B., Bousquet F., Mathevet R., 2021. Serious Games, In: Biggs R., de Vos A., Preiser R., Clements H., Maciejewski K., Schlüter M. (eds), *The Routledge Handbook of Research Methods for Socio-Ecological Systems*, Taylor & Francis, 176-188. https://doi.org/10.4324/9781003021339-15.

Becu N., 2020. *Les courants d'influence et la pratique de la simulation participative : contours, design et contributions aux changements sociétaux et organisationnels dans les territoires,* Habilitation à diriger des recherches, La Rochelle Université.

Becu N., 2022. Usability of Computerised Gaming Simulation for Experiential Learning, *In:* Castro L. M., Cabrero D., Heimgärtner R. (eds.), *Software Usability*, IntechOpen, https://doi.org/10.5772/intechopen.97303.

D'Aquino P., Bah A., 2013. A Participatory Modeling Process to Capture Indigenous Ways of Adaptability to Uncertainty: Outputs From an Experiment in West African Drylands, *Ecology and Society*, 18 (4). https://doi.org/10.5751/ES-05876-180416.

D'Aquino P., 2016. TerriStories, un jeu au service de l'invention collective dans les politiques publiques, *Animation, territoires et pratiques socioculturelles*, 10, 71-80.

Duke R.D., 1974. *Gaming: The Future's Language*, London: SAGE.

Duke R.D., 2011. Origin and Evolution of Policy Simulation: A Personal Journey, *Simulation and Gaming*, vol. 42, n° 3, p. 342-358. https://doi.org/10.1177/1046878110367570.

Hassenforder E., Barreteau O., Daniell K.A., Pittock J., Ferrand N., 2015. Drivers of Environmental Institutional Dynamics in Decentralized African Countries, *Environmental Management*, 56(6), p. 1428-1447. https://doi.org/10.1007/s00267-015-0581-2.

Hassenforder E., Clavreul D., Akhmouch A., Ferrand N., 2019. What's the Middle Ground? Institutionalized vs. Emerging Water-Related Stakeholder Engagement Processes, *International Journal of Water Resources Development*, 35 (3): 525-42. https://doi.org/10.1080/07900627.2018.14 52722.

Klabbers J.H.G., 2009. *The Magic Circle: Principles of Gaming & Simulation.* Rotterdam, The Netherlands: Sens publishers. https://doi.org/10.1111/j.1467-8535.2007.00725_5.x.

Mathevet R., Bousquet F., 2014. *Résilience & Environnement, Penser Les Changements Socio-Écologiques,* Buchet Chastel.

Ostrom E., 1990. *Governing the Commons: The Evolutions of Institutions for Collective Action,* Cambridge University Press.

Insert 2

LittoWAG: a companion game for coastal risk adaptation

Julie Latune, Mariana Rios, Eva Perrier, Geraldine Abrami and Nils Ferrand

LittoWAG is a role-playing board game developed in the LittoPart project, part of the Innovation program "Littoral+" funded by the Banque des Territoires and the Occitanie Region, within which it aims at tackling the "collective and citizen resilience" target.

The game, inspired by Wat-A-Game, has been designed through:
– interviews with citizens along the Occitanie coastline,
– design and testing in the laboratory,
– validation testing in workshops with citizens and experts.

Primarily aimed at citizens and users of the coastline, its objective is, after a discovery and sensitisation phase, to collect the perception of citizens on the management and adaptation of the coastline to risks. The tool is engaging and it helps citizens grasping the risks of erosion and marine submersion. The results will be used to feed the reflections for local and regional strategies.

The players must avoid ending the game with the feet wet, thanks to individual actions (economic activities, housing repairs, moving, etc.) and collective actions of coastline management (hard management like dikes or groins, nature-based solutions, adaptation, reshaping of the space). The hazards and the choices of the actors affect stakes' indicators (tourist economy, biodiversity, agriculture, public services, real estate heritage, well-being).

Figure I2.1. The LittoWAG game board

This game is currently being deployed, and the first workshops have produced the following initial results, addressed on the base of the ENCORE framework (chapter 18): It has been used (2023) to support a citizen panel building an alternative strategic action plan for coastal area management in the French Occitanie region.

Learnings	Cognitive	"The game makes people aware of the complexity of all the existing constraints: tourism, ecosystems, local elected officials, agriculture ..." "There is no good solution", "Any action can have a beneficial and harmful influence"
	Normative	"The game shows the different possible management alternatives to be implemented: hard management, soft management, and or adaptation." "We are obliged to protect everyone otherwise those who are not protected will say: why him and not me ? "
	Relational	" Long game but has the merit to exchange between interlocutors with different or even opposing interests."
Perceptions	Coastline solutions	"Better [...] to reconstitute the dune than to put sand back [only] for the tourists." "The recharging of sand is without interests".
	Priorities	"The most at stake are where there are houses."
	Land use changes	"It is either the urbanised parcels or the agricultural parcels that have been transformed. These parcels have become respectively parcels of natural zones and parcels of urban zones". The territory has lost agricultural surface (but natural spaces are preserved). Urban areas of the seaside are replaced by natural areas to recreate a space of evolution for the coastline.

Chapter 13

"L'Eau en Têt": An educational role-playing kit to educate people on participation

Patrice Robin, Fabrice Carol, Floriane Le Moing, Géraldine Abrami,
Nils Ferrand and Dominique Dalbin

Starting in 2014, a multi-partner project involving an agricultural school, a research team and a watershed syndicate from the Eastern Pyrenees region of France led to the design and publication of "L'Eau en Têt"[1], a role-playing game devoted to sustainable water management. This chapter presents the initiative, reviews the pedagogical and other uses of this game kit, and addresses which learning effects this type of approach has on participants.

Territorial issues such as water management or, more broadly, the management of common goods in rural and suburban areas, are central to education in the agricultural sector in France (Castel, 2014). Beyond the pedagogical methods used in their teaching, these issues are also at the heart of the facilitation and development mission of the regions assigned to the Local public establishments for agricultural education and vocational training (EPLEFPA[2]).

A serious game is defined as a device combining a pedagogical scenario and a game (Djaouti *et al.*, 2017). The use of serious games in the classroom is on the rise and valued in agricultural education, particularly simulation games in the field of planning and agriculture, as they allow complex situations to be tackled in an interdisciplinary manner. They allow for experimentation of hypothetical scenarios in a protected environment, for learning by trial and error through problem-solving exercises, and for participants to familiarise themselves with modelling devices (Inspection de l'enseignement agricole, 2019; Rouchier, 2018).

The role-plays presented in the chapter 13 are a particular type of simulation game.

Developed through research for use in participatory approaches around the management of common goods (Rouchier, 2018), this role-playing approach also lends itself well to use in a school setting. In this context, a teacher who was trained in using the Wat-A-Game kit (see box 13.1) developed by the G-EAU joint research unit "Water Matters" in Montpellier, began creating role-playing games in 2012 on agricultural water sharing and using them in class at the EPLEFPA in Perpignan-Roussillon. Based on this

---

1. L'Eau en Têt is a word play in French which could be translated by "Water at the helm".
2. Établissement publics locaux d'enseignement et de formation professionnelle agricole.

experience, starting in 2014, the EPLEFPA in Perpignan-Roussillon, the Têt watershed syndicate (SMTBV[3]) and the G-EAU joint research unit embarked on a partnership project to design and publish an educational role-playing game dedicated to sustainable water management and intended primarily for a high school and student audience[4]. L'Eau en Têt was developed at EPLEFPA with the support of INRAE researchers in using the Wat-A-Game design methodology for the Têt catchment area. The SMTBV provided data on existing uses and their functioning, and participated in the calibration of the game in order to make it realistic and adapted to the Têt river context.

▸▸ L'Eau en Têt: an educational role-playing kit for sustainable water management

L'Eau en Têt was designed to address the issues of quantitative management, i.e. the question of sharing surface and underground water resources in order to guarantee the flow necessary to the proper functioning of watercourses while meeting human uses. For this purpose, three game boards were developed covering three different socio-geographical scales linked to the Têt catchment area: the entire catchment area, the area downstream of a dam (corresponding to the Roussillon plain located downstream of the Vinça dam) and a local irrigated area illustrating the situation of an authorised syndicate association (ASA), in charge of managing a network of irrigation canals[5] (figure 13.1).

In a complementary way, the fourth game board focuses on the qualitative management of water through the risks of diffuse pollution linked to the use of herbicides.

Students are in charge of a specific role which can be, depending on the game board chosen, a farmer, a local authority, a dam or a canal manager, an industrialist or a local citizen. Through several rounds of the game, which can represent, for example, an irrigation season, a summer or a whole year, the students encounter a certain number of more or less problematic situations that they have to face alone and in groups, by developing strategies that can be more or less cooperative. To carry out these strategies, they can implement actions that allow them to initiate change, notably in the activities carried out, the use of water resources or the existing equipment and infrastructure.

▸▸ L'Eau en Têt: educational uses in agricultural high schools

L'Eau en Têt offers a rich and modular game base that can be used to meet a variety of learning objectives, as illustrated by the diversity of educational situations in which it has been used in agricultural schools since its first use on the Théza site in 2016-2017.

A first learning objective may simply be the acquisition of knowledge and know-how related to the management of water as a common resource in a territory. The agricultural school LEGTA[6] in Théza offers a teaching module devoted to sustainable

---

3. Syndicat mixte de la Têt Bassin Versant.
4. This project was financed by the Rhône-Mediterranean-Corsica water agency, the Occitan Pyrenees-Mediterranean Regional Council and the European Agricultural Fund for Rural Development.
5. On the scale of the Têt watershed, these ASAs manage a network of irrigation canals equivalent to around eight times the length of the river.
6. Lycée d'enseignement général et technologique agricole.

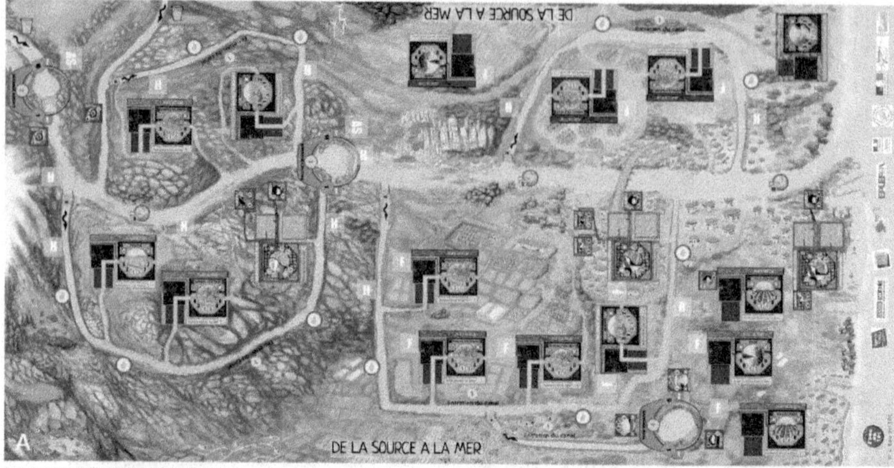

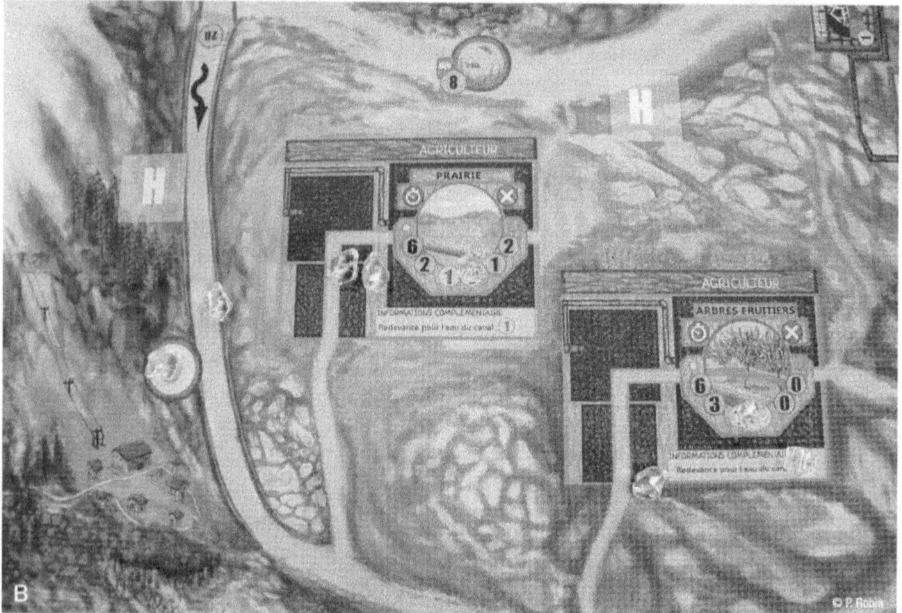

Figure 13.1. One of the three L'Eau en Têt game boards (A) and details (B)

water management in the Mediterranean context in the scope of their "Horticultural Production" certification programme. L'Eau en Têt is used in this module, in conjunction with meetings with local stakeholders organised in partnership with the SMTBV, as a way of confronting the concrete problems of water management in general, and of this catchment area in particular.

A further objective is the acquisition of professional skills. Since 2019, for example, teachers at the vocational agricultural high school in Rivesaltes (Eastern Pyrenees) have been experimenting with the use of L'Eau en Têt with a vocational baccalaureate class titled "Horticultural Production Management", as the recent reform of the vocational agricultural baccalaureate has highlighted the notion of common resources.

In these sessions, the game is explicitly introduced as a means of placing the students in a professional situation that could be their own in the future, for example as an irrigating farmer.

Finally, within the scope of agricultural education, both general and technological training, L'Eau en Têt can be used to address the issues of sustainable territorial development of rural and suburban areas and also to complement students' methodological skills and knowledge necessary for their future studies. In this way, L'Eau en Têt can be used in the preparatory phase for a territorial study to better assimilate the notion of common goods, territorial stakes and the methods to manage these common goods, illustrated through the example of water. In this context, the game board represents a territory in which the actors, i.e. the student players, are confronted with a problem that they have to try to solve. Through the use of modelling, L'Eau en Têt thus enables students to effectively tackle relatively complex issues that may otherwise be very abstract to them, as illustrated in box 13.1.

> **Box 13.1. Testimony from the Bourg-en-Bresse public school: when students design a role-playing kit or modelling as a learning approach**
>
> In 2019-2020, as part of a national project dedicated to the promotion of water-related projects in agricultural education, a team of teachers working with year 11 students in the ecology, agriculture, territory and sustainable development programme in Bourg-en-Bresse built a course in partnership with a local watershed syndicate. Allowing the students to build the game kit themselves, and based on the experience of L'Eau en Têt, the objective was to model how the catchment area functions. To build the game kit, water uses and the stakeholders needed to be identified and a certain issue had to be chosen, for instance a specific pollution problem, in order to model the situation. This project was based on a mission statement that was drawn up in collaboration with the catchment syndicate. The game kit would later serve as a support to facilitate workshops carried out by the syndicate. The COVID-19 crisis interrupted the project, but an initial prototype of the game kit had been developed and plans to continue the project have been laid out.

▶▶ What did the students learn from L'Eau en Têt?

Consistent evaluation of learning requires the development and deployment of an elaborate monitoring and analysis system (see chapter 12). Within the scope of this project, elements for evaluation could only be collected at the end of the session, through questionnaires on the ten-some sessions conducted between 2016 and 2019 in four different establishments as well as through two series of filmed interviews conducted in 2017 and 2019 at the LEGTA in Théza. Depending on the situation, 65% to 90% of the students found their involvement in the game satisfactory and the experience interesting. The teachers noted that the students' personal involvement in the session favoured long-term memorisation, which could then be used in subsequent lessons.

Two constitutive aspects of this type of role-play help reinforce involvement and memorisation. The immersive nature of the role-playing session requires personal commitment to a specific role in a particular setting. Strong interactions with other students activate an affective dimension which creates a strong impression that becomes memorable and which can be subsequently re-activated. Many studies devoted to serious games highlight the importance of this experiential dimension and the interaction between players in the learning process (Plass *et al.*, 2015; Bado, 2019).

In addition, throughout the game, feedback provided in the form of a dashboard on the evolution of their economic situation, that of the other players and the state of the resource, allows each player to reflect on their choices and contribution to the group (Lavoué, 2012). Furthermore, by explicitly focusing the attention of learners on the pedagogical objectives of the session and by reinforcing learning through targeted debriefing, the effectiveness of the role-play is enhanced (Becu, 2020).

Lastly, in the context of a participatory process, the use of a role-playing game must be part of a well thought-out and coherent whole (see chapter 13). An educational role-playing game such as L'Eau en Têt can therefore only be effective in terms of learning if it is part of a coherent pedagogical process built by the teacher based on the learning objectives (Djaouti *et al.*, 2017; Inspection de l'enseignement agricole, 2019).

More specifically, in terms of learning, 90% of respondents felt they had learned something. If we take the categories defined in the article by Abrami *et al.* (2019), the learning most frequently cited in the questionnaires is cognitive (concept of common resources, actors in water management, interdependencies between actors and between actors and resources, specific concepts such as gross and net abstraction, biophysical processes, alternative solutions to herbicides, etc.), and encompasses the learning of social skills (discussion, participation in debate, development of a critical mind). Normative or affective learning took place as well (the need for coordination between actors, the need to develop responsible attitudes). This is particularly well highlighted in the filmed interviews (“*We had to see with our colleagues to get along*”; “*We put our cards together*”; “*We saw several ways of thinking*”).

▸▸ What are the prospects for the deployment of L'Eau en Têt?

The L'Eau en Têt game kit was used in other regions beyond the Têt watershed in 2017 (box 13.2) and can now be further distributed to other agricultural schools.

As these experiences show, L'Eau en Têt is a rich and modular tool, but it is complex to grasp. The development of “ready-to-use” tools along with specific training and objectives could help make it easier for teachers and trainers, who did not partake in their design and who are not familiar with Wat-A-Game, to appropriate them. Moreover, if the learning experience is to be further detailed, dedicated monitoring and evaluation methods need to be designed and implemented. The chapter 12 provides insight into the challenges of designing such methods and proposes a framework for evaluating learning as applied to participatory approaches. Finally, although bringing L'Eau en Têt into agricultural programmes is a central objective, making this game kit available for other activities conducted with external partners remains a significant objective for EPLEFPA Perpignan-Roussillon. A few experiments have been conducted in the Eastern Pyrenees so that local actors and elected officials in catchment areas can get to know and use the game kit. However, as L'Eau en Têt was designed mainly for educational purposes, it seems less suited to supporting decision-making in real situations. Other uses could be foreseen, but here too it would be necessary to simplify the game and develop adapted sessions based on existing work on other games (Dernat *et al.*, 2021).

Box 13.2. Testimony from the Higher Institute for Agricultural Studies – ENSFEA in Toulouse

In 2017, three trainee teachers tested the game kit in their traineeship at ENSFEA* in Toulouse. These test sessions were conducted in their respective establishments spread over the metropolitan area. Although L'Eau en Têt was designed for a specific watershed, it can be used in other contexts.

"As part of our year-long traineeship in agricultural education, we took part in testing the 'L'Eau en Têt' game kit for water management. The game was tested in six sessions with our classes. The objective being the evaluation of the pedagogical added-value, we evaluated these sessions through our observations and using questionnaires distributed to the participants: students and teachers. The conclusions are mixed at this stage of the experiment. On the one hand, the majority of the students had a positive experience, and the questionnaires and debriefings showed that the targeted problems were well integrated. The fact that it was a role-playing game, that placed them in a real situation, as well as the realism of the game, can be retained as the main factors. However, the teachers and some of the students found the rules too complex, making the game difficult to grasp. Recommendations for improving the game and its use include the standardisation and simplification of the rules and the organisation of training sessions for users. In the field of economic, social and management sciences, it seems that the game can be used in many modules, both general and professional, in particular to question the logic of actors and their relations to and within a territory" (Marie Guérin, Kevin Cuevas and Nathalie Billot, teachers).

* École nationale supérieure de formation de l'enseignement agricole

⇥ L'Eau en Têt: a learning tool on participation for regional actors of tomorrow

Education in the agricultural sector aims to train future regional actors with the necessary skills for sustainable development. If development processes are to be more and more based on participatory territorial approaches in the future, it seems essential to encourage teaching methods that make future citizens aware of these approaches and equip them with the skills to take part in them.

An interesting example is the Pollution-Solutions educational role-playing game (Rouchier, 2018). It is the same type of tool as L'Eau en Têt (multi-player simulations in the form of role-playing games), which was designed to work in a school setting on the issue of managing a common good. Using game-based learning to experiment the tension between a common good (air quality for instance) and individual interests, the authors argue that Pollution-Solutions allows students to recognise this tension, and thus to take the first step in learning to reflect on it and manage it in their future life as citizens.

For Becu (2020), three types of learning are strengthened through role-playing designed with participation in mind, thus increasing actors' participatory skills: political/communicative learning (learning to debate, deliberate), cognitive learning (building systems thinking on the territory), and organisational learning in view of guiding collective action. In the cases of Pollution-Solutions and Eau en Têt, the first two types of learning are present since the students have to analyse situations collectively, set common goals, then propose and implement actions. Furthermore, Eau en Têt provides students with an opportunity to familiarise themselves with a tool for participatory processes and participatory modelling. Perhaps it even spurs them into thinking about their professional lives or provides them with their calling.

Lastly, we conclude with a few points from Rouchier (2018). Such tools may well contribute to educating for participation and, more generally, for citizenship and sustainable development, especially as the school setting encourages the acceptance of new ideas. However, to be effective, such experiences must be continually repeated over the formative years, need to be integrated into ambitious and coherent educational programmes as well as properly assessed.

▸▸ References

Abrami G., Aucante M., Ducrot R., Ferrand N., Hassenforder E., Noury B. *et al.*, 2019. Formations à la construction collective de jeux : types d'apprentissages et place des supports numériques. Conference_item. s.n. 2019. https://agritrop.cirad.fr/592903/

Bado N., 2019. Game-Based Learning Pedagogy: A Review of the Literature, *Interactive Learning Environments* 0 (0): 1-13. https://doi.org/10.1080/10494820.2019.1683587

Becu N., 2020. *Les courants d'influence et la pratique de la simulation participative : contours, design et contributions aux changements sociétaux et organisationnels dans les territoires*, Habilitation à diriger des recherches, La Rochelle Université. https://hal.archives-ouvertes.fr/tel-02515352

Castel P., 2014. L'ancrage territorial du lycée agricole : une spécificité au service d'une pédagogie innovante de la complexité et de l'action, propice à l'éducation au développement durable (EDD). https://halshs.archives-ouvertes.fr/halshs-01199629

Dernat S., Rigolot C., Vollet D., Cayre P., Dumont B., 2021. Knowledge sharing in practice: a game-based methodology to increase farmers' engagement in a common vision for a cheese PDO union, *Journal of Agricultural Education and Extension*, 28:2, 141-162. https://doi.org/10.1080/1389 224X.2021.1873155

Djaouti D., Alvarez J., Rampnoux O., 2017. *Apprendre avec les serious games ?*, Éclairer, Réseau Canopé. https://hal.archives-ouvertes.fr/hal-02533902

Inspection de l'enseignement agricole, 2019. *Les jeux sérieux : intérêts pour la formation des apprenants et l'enseignement agricole ?*, Rapport Général 2017-2018, Ministère de l'Agriculture.

Lavoué E., 2012. *Towards Social Learning Games*, 11th International Conference on Web-Based Learning (ICWL 2012), 168-77, Lecture Notes in Computer Science, Sinaia, Romania, Springer, https://hal.archives-ouvertes.fr/hal-00731093

Plass J., Homer B., Kinzer C., 2015. Foundations of Game-Based Learning, *Educational Psychologist*, 50, 258-83. https://doi.org/10.1080/00461520.2015.1122533

Rouchier J., 2018. Les serious games et l'éducation au bien commun : L'exemple du jeu PollutionSolutions, *Action Publique - Recherche et Pratique*. https://hal.archives-ouvertes.fr/hal-02067357

Integrating expert technical knowledge in a sanitation planning process with stakeholders: WasteWAG modelling and game in Senegal

Mélaine Aucante, Rémi Lombard-Latune, Alpha Ba, Camille Cheval, Paul Moretti and Nils Ferrand

WasteWAG (wastewater game) is a role-playing game and participatory planning tool for individual and collective sanitation systems designed for urban and rural areas of Senegal. It was modelled over several successive stages alternating prototype production, testing in controlled conditions (by researchers) and testing in the real environment (with civil society organisations and field actors). This chapter aims to present the singularity of this modelling process, which contributed to the transfer of technical knowledge to local stakeholders.

▸▸ Sanitation planning, a far cry from traditional recipies of development

Access to sanitation for all, and development of the services that support it, is a major issue as highlighted by the Sustainable Development Goal (SDG) 6.2: "By 2030, achieve access to adequate and equitable sanitation and hygiene for all and end open defecation" (UNDP, 2015). Construction of sanitation services, whether on the scale of a large city or on a more local level, is sensible. The public actor legally competent in the matter (local authority or the state if the competence is not decentralised) commits to this aim beyond its own responsibility, as the end users directly (fee), or indirectly (public funds) ensure its financing. Sanitation needs analysis and the design and choice of technical response are undertaken in the sanitation project's planning phase, which generally corresponds to the drafting of a sanitation framework plan (SFP). The SFP's feasibility study is generally entrusted to consulting firms that sometimes try to analyse the needs from the users' standpoint by using surveys. More often, users are defined in socio-economic indicators for project and infrastructure sizing. Other stakeholders, such as existing technical services, which will be responsible for operating the facilities, are not systematically involved.

Investing in the sanitation planning phase of a project is particularly interesting for the sustainability and quality of sanitation services. The assumption is the following: by consulting all involved stakeholders in the planning phase, the future sanitation service will be better suited to the reality and stakeholders will understand and endorse their role. Indeed, it is the local actors themselves who are in the best position to establish a balance between their needs/constraints and the means they can allocate to sanitation. Their participation also provides them with a better understanding of their roles in the service. In order for the greatest number of stakeholders to be able to contribute to these choices, they first need to be provided with knowledge, as well as an overview of the issue based on key stakes: financial constraints (investment in and maintenance of facilities), individual and collective infrastructure requirements, as well as the technical expertise required to implement and maintain the facilities. This concerted planning is not intended to replace technical studies, but rather to complement them by providing orientation, to ensure the support of local actors and the sustainability of the project.

Users can participate in different design stages through a variety of approaches. The Companion Modelling and the CoOPLAGE toolset offer methods for varying degrees of participation, using several tools with multiple functionalities. It is worth questioning to what extent the development and application of a multi-functional tool requiring active users' participation allows for the planning of coherent sanitation systems.

INRAE research teams (Research units G-EAU in Montpellier and REVERSAAL[1] in Lyon) have been working together since 2015 on constructing a concerted sanitation planning approach. It was tested for the first time in Senegal (figure 14.1), in urban and rural areas, in the scope of PLANISSIM[2] which was backed by the non-governmental organisation ACTED in 2017-2018.

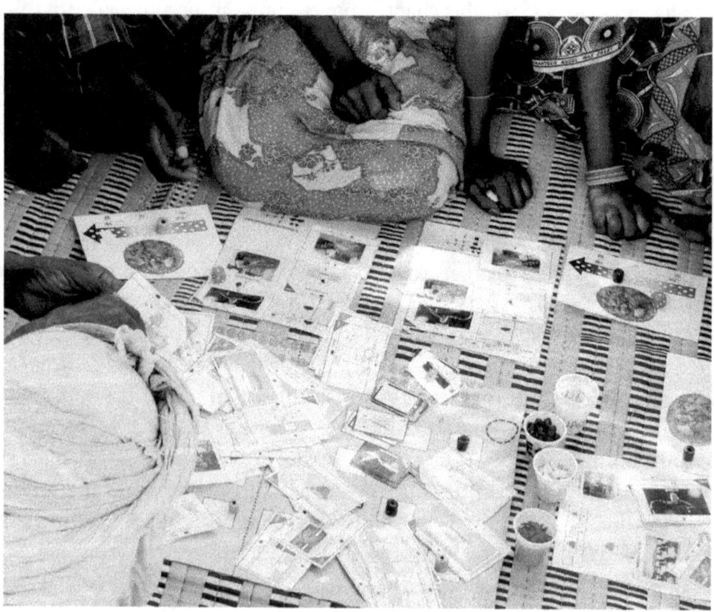

Figure 14.1. Awareness-raising workshop involving WasteWAG in the Ranérou department

---

1. REVERSAAL: Research unit (Reduce, reuse, recover wastewater resources).
2. PLANISSIM (PLanification de l'AssaiNIssement par modéliSation et SIMulation participative): Planning sanitation through modelling and participatory simulation.

▸▸ The PLANISSIM project: where social engineering and process engineering meet - Building an interdisciplinary approach to sanitation

Sanitation planning, as mentioned above, poses a number of cognitive, social and methodological constraints. Cognitive because it deals with complex technical objects on which the participants have varying degrees of knowledge. Social because sanitation planning prefigures the sanitation service, i.e. the pooling together of different resources by a set of actors with different interests, needs and objectives. Methodological because it requires the integration of information from multiple disciplinary fields (hydrogeology, process engineering, economics, sociology, law, etc.) and must allow the participants to make coherent proposals. Sanitation planning must integrate all this complexity. It was therefore decided to accompany the actors through an interdisciplinary posture between social engineering and process engineering by using multi-stakeholder and multi-level participation methods.

Participation engineering to build and equip the process

According to Arnstein (1969), there are many ways of encouraging the participation of service users who are involved at different levels: information, consultation (gathering opinions), co-construction (dialogue, co-constructing proposals), decision (deciding together, giving the power to decide) (see chapter 2 and 18). Companion Modelling is the posture adopted here (Etienne, 2010), which in theory is situated at a co-construction level.

Developed by the ComMod collective, Companion Modelling is a multi-actor process aimed at modelling "complex and dynamic study objects that are also objects of multiple actions and stakes" (ComMod Collective, 2004). The objective of this approach is to build a model, as a shared representation of reality by including the standpoints and knowledge of the various participants. This model is not only the result of a collective effort that focuses in particular on the visions and constraints of the actors with regard to the system under study, but it can also serve as an interface between them in the form of a role-playing game or a simulation tool. It can thus be used to explore different scenarios for modifying the system under study and to discuss their impacts.

This approach was associated with the one developed in the CoOPLAGE set of tools and methods to assist stakeholders in the design, implementation and monitoring-evaluation of their participatory process (Ferrand *et al.*, 2016). Multi-actor and multi-level, they allow actors of various nature and responsibility to take part in the same decision-making process. Two methods in particular were mobilised: Wat-A-Game (WAG) to model sanitation systems, and CoOPLAN for planning purposes. They resulted in the creation of an approach halfway between a role-playing game and a planning matrix (see chapters 2, 13 and 18).

Companion Modelling and the CoOPLAGE approach have many points in common for model development, in particular through the development of conceptual diagrams, as formalised in the ARDI (Acteurs, Ressources, Dynamiques, Interactions; Etienne *et al.*, 2011) or WAG method.

A particular feature of the CoOPLAGE approach reflected in its acronym- is the desire to empower participants. This is achieved through the use of a modelling

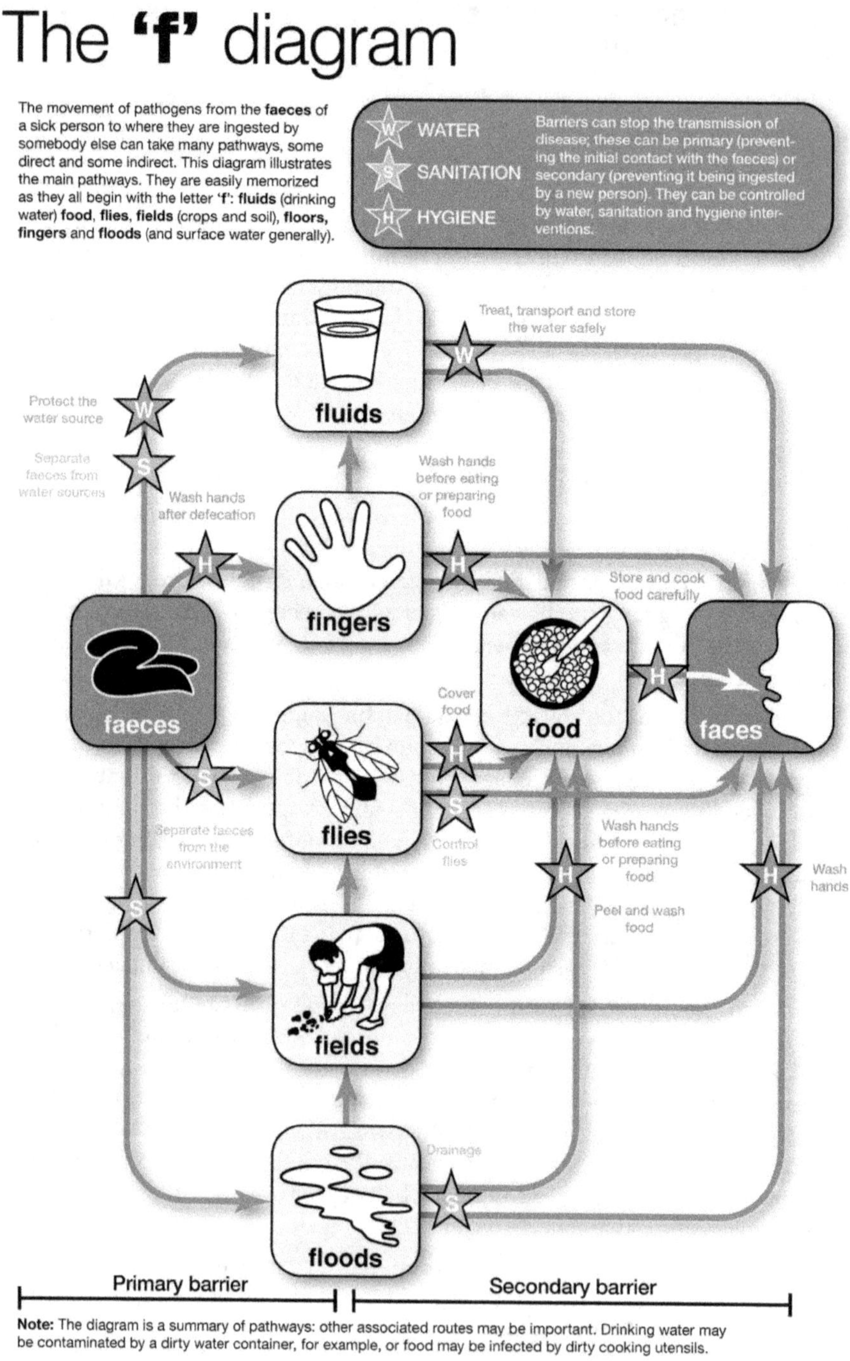

The 'f' diagram
The movement of pathogens from the faeces of a sick person to where they are ingested by somebody else can take many pathways, some direct and some indirect. This diagram illustrates the main pathways. They are easily memorized as they all begin with the letter 'f': fluids (drinking water) food, flies, fields (crops and soil), floors, fingers and floods (and surface water generally).
W WATER
S SANITATION
H HYGIENE
Barriers can stop the transmission of disease; these can be primary (preventing the initial contact with the faeces) or secondary (preventing it being ingested by a new person). They can be controlled by water, sanitation and hygiene interventions.
fluids
Treat, transport and store the water safely
Protect the water source
Separate faeces from water sources
Wash hands after defecation
Wash hands before eating or preparing food
fingers
Store and cook food carefully
faeces
food
faces
Cover food
flies
Separate faeces from the environment
Control flies
Wash hands before eating or preparing food
Peel and wash food
Wash hands
fields
Drainage
floods
Primary barrier
Secondary barrier
Note: The diagram is a summary of pathways: other associated routes may be important. Drinking water may be contaminated by a dirty water container, for example, or food may be infected by dirty cooking utensils.

Figure 14.2. Fecal-oral contamination pathway diagram (source: WEDC, 2014)

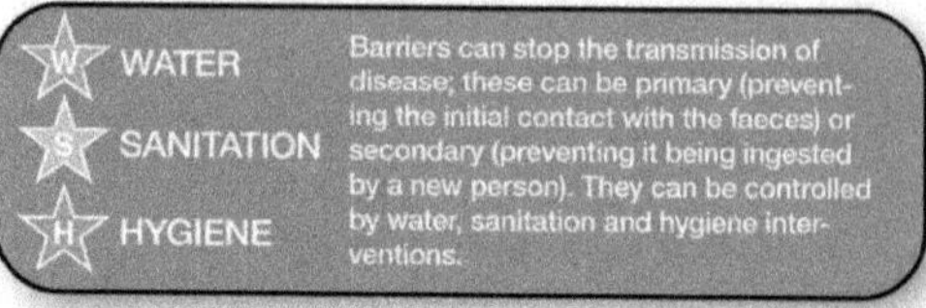

method leading to the production of simple models[3] to represent complex systems, and above all through the transfer of the tools produced and the basic skills needed to use them. The PLANISSIM project is notably based on this empowerment dimension. It is innovative in having created a multifunctional tool (awareness raising, planning, simulation) with a unique grammar, thus bridging two until now distinct methods (WAG and CoOPLAN), by in particular including expert technical knowledge on sanitation technologies.

Process engineering for sanitation as a source of technical expertise

As stated in our initial hypothesis, the planning of sanitation systems requires the mobilisation of technical knowledge. As this knowledge is not prominent or at least not shared among the actors, a number of key concepts on sanitation have been integrated into designing the model:
– the fecal-oral contamination pathways widely described by development aid (figure 14.2) have been integrated to highlight the links between sanitation and population health;
– the concept of the sanitation chain (figure 14.3), which demonstrates that sanitation is an orderly sequence of technical devices, which allows the handling and management of materials from their production to their reuse or disposal (Gabert *et al.*, 2010). This representation through individual links makes it possible to break down the problem by asking the right questions at the right scale (individual and collective). Depending on the context (urban and rural in particular), the number of links may vary but the principle remains the same.

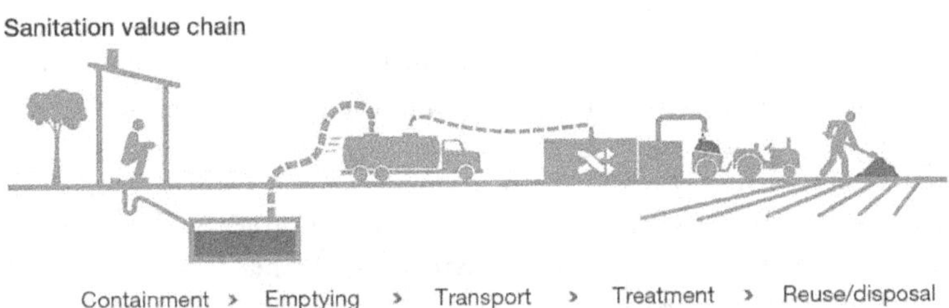

Figure 14.3. The sanitation value chain

These two elements were integrated into the conceptual diagrams used in the laboratory modelling phase, which produced the framework for the model: the variables and elements that are to be mobilised by the participants, as well as the indicators that make it possible to characterise these elements and to monitor the evolution of the situation during the simulation phase of the tool that is to be created.

Subsequently, data had to be collected on the field to identify the local variations of each of the two concepts: main contamination routes and available technologies for each of the sanitation links. The characterisation of needs and constraints present on

---

3. In modelling jargon, different types of models are distinguished in function of their characteristic. For example, 'KISS' (Keep It Simple Stupid!), as opposed to 'KIDS' (Keep It Descriptive Stupid!). Here, we opted for a KILT approach (Keep It a Learning Tool!), with reference to the model's aim, as proposed by Christophe Le Page (2017).

the field was also necessary to identify the relevant technical options to be added to those already present in the model in order to propose adapted innovations capable of making the situation evolve.

Finally, to enrich the discussions and the range of technical possibilities, a database of 25 sanitation technologies is used in the model. It was built from a compilation of bibliographical references (in particular the Sanitation Compendium; Tilley *et al.*, 2014) and an analysis of the existing situation in the field carried out during the first months of the project. The various technologies are presented by diagrams and a few indicators on the game cards. They are also detailed on a double-page spread in a booklet made available to participants, and are the subject of an oral presentation by experts during planning workshops.

The interdisciplinary approach described above requires a certain anchoring in the study areas: need for technical knowledge, participation of users and experts. For this purpose, a close collaboration was carried out with ACTED, based in Senegal, and members of Senegalese civil society organisations (CSOs).

In the following section, we will see how these partnerships have allowed us to build and apply a multifunctional tool adapted to different contexts, as well as its main results.

▶ WasteWAG construction and application: a multifunctional tool for planning wastewater systems

Participants in the modelling process and facilitators on the field: CSO members play a central role

Very early on, ACTED identified members of CSOs who could participate in the modelling process and facilitate the approach at the field level within each intervention zone. A variety of profiles were identified: representatives of neighbourhood committees, municipal employees, Senegalese or international NGOs, etc. Interventions took place at the urban level (in city neighbourhoods) or at the rural level (in villages), within the participants' direct environment, which allowed them to better appropriate the studied object.

As the facilitators were not experts in sanitation issues and technologies, they were trained so that the approach could be well implemented. In addition to being partners in the modelling process, they were also trained in the tool construction phases. These elements are detailed in table 14.1.

The prior knowledge of CSO members, coupled with their training in modelling and the application of WasteWAG in the field, helped to anchor the planning process in an operational dynamic. Their input was invaluable at every stage in constructing the final tool.

A tool for planning, but which tool?

WasteWAG (for wastewater game) is the name given to the tool resulting from the modelling process largely carried out during the PLANISSIM project. Its modelling was carried out in several successive stages, alternating between the production of prototypes, testing of the tool under controlled conditions (among research actors) and testing in real conditions (with CSOs and other field actors).

Table 14.1. Content of the collective modelling phases and training of civil society organisations (CSOs).

Workshops	Initial training of facilitators	Awareness raising	Planning	Simulation
Content	1) Participatory processes and facilitation	Highlighting the impacts of practices on health, whether individual or collective. Fecal-oral contamination routes	1) Technical matrix: constitution of sanitation chains	1) Testing of the different scenarios produced on the basis of the planning workshop
	2) Basic notions of sanitation and selected technologies		2) Social matrix: what organisation should be put in place to design, implement and maintain the pre-defined chain?	2) Amendment of scenarios (new planning phase)
	3) Development, use and facilitation of WasteWAG		3) Cost matrix: summary of the various costs and their distribution among the actors	3) Testing the scenarios
Duration	3 days	1/2 day	2 days	1 day
Participants	Trainers, facilitators	Facilitators, users	Facilitators, users	Facilitators, users

Our starting hypothesis allowed us to draw up the following "specifications": Our tool had to be didactic (accessible to the general public), progressive and multifunctional (several steps to assimilate the complexity and knowledge necessary to make informed choices), contain synthesised and contextualised technical data, and finally be simple enough so that its facilitation does not require expertise on sanitation and participatory processes.

On the basis of common grammar elements from the Wat-A-Game method, the foundations of the WasteWAG model were laid out in the laboratory: spatial environment, key concepts of sanitation, essential resources and dynamics, actors. Feedback from citizens and CSOs collected during exploratory field visits and a launching workshop were also added (perceptions of water flows, terms and representations used, users' place in the sanitation service, etc.).

WasteWAG as an awareness-raising tool

WasteWAG takes the form of a board game in which participants take the role of heads of households (figure 14.4A). They have to manage their sanitation according to different technical solutions represented on cards (figure 14.4B). Materials (raw or treated) that are not properly managed end up on the collective game board (figure 14.4C, neighbourhood or village depending on the area). These generate flies (representing indirect contamination routes, see figure 14.2), and contaminate the family meal (the famous Senegalese *Tiep Bou Dien*), resulting in a decline in the health of the household.

WasteWAG as a planning tool

Planning in WasteWAG is approached using two matrices, technical and social (figure 14.4E). The technical matrix shows the different links in the sanitation chain (see figure 14.3), which the participants fill in with the technical solution cards. For this new phase, key elements have been added (figure 14.4D): incoming and outgoing resources, place in the sanitation chain, construction and maintenance costs, space and time requirements, resources needed to build, possible valuable products, etc. The proposals are then discussed in groups in order to choose the system(s) best suited to the local constraints.

The social matrix leads each group to reflect on the means necessary at different temporal and spatial scales to implement the chosen system. For this purpose, a non-exhaustive set of individual or collective action cards were created, such as: "Request a permit from the local authorities", "Organise a meeting", "Sell my livestock", etc.

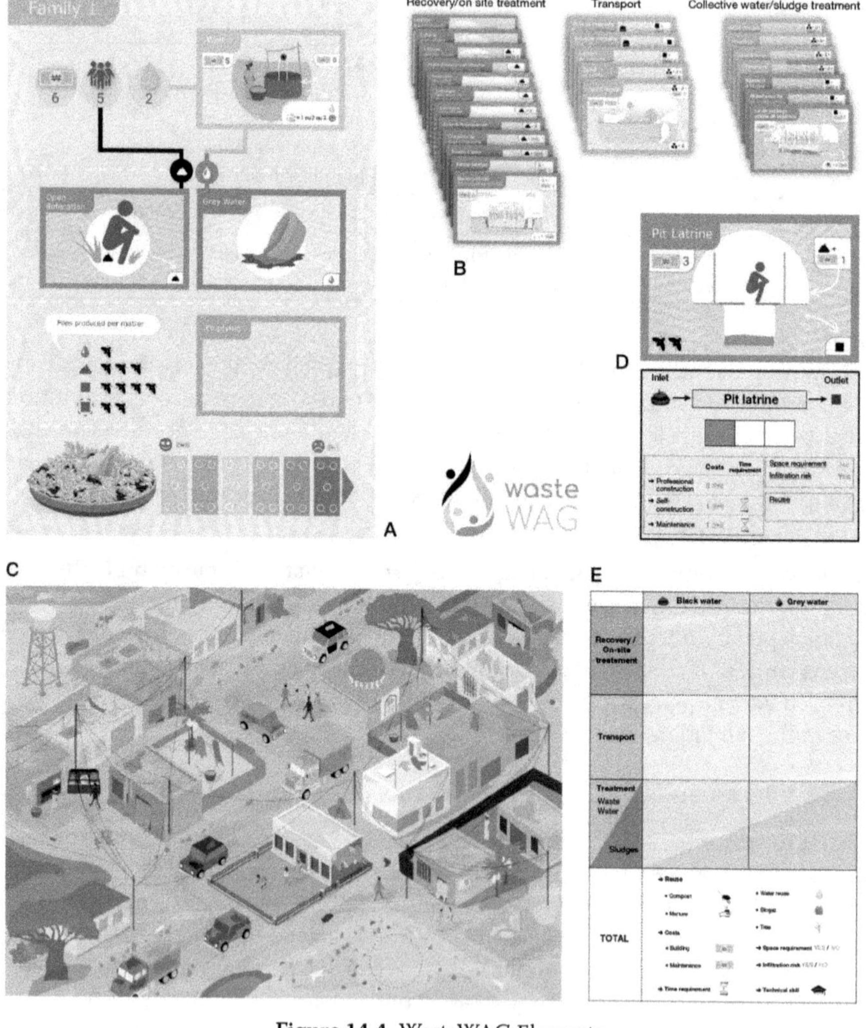

Figure 14.4. WasteWAG Elements

Finally, where there was sufficient time, a cost matrix was added. It provides a summary of the different costs (capital and operating) of the different proposals.

WasteWAG as a simulation tool

In the last working phase, the project team analysed the different sanitation systems produced and drew a conclusion in the form of three scenarios for each of the intervention areas, based on the frequency of appearance of the different technical solutions in the different proposals, as well as their technical, economic and social viability within the context of each area. During the final simulation phase, which took place in the summer of 2019, the scenarios were tested on the model/board game by the participants to assess their feasibility and impact, before being reworked to improve them.

And what does this mean in practice?

The use of WasteWAG in the field was carried out in several phases, as was its design. The content of the different sessions is presented in table 14.1. The number of participants in the different stages of the project and some of the results are shown in figure 14.5.

At the end of the first phase, the model is considered realistic and appropriate for launching a consultation process on the subject of sanitation. The second planning stage also shows some success. However, whereas the technical planning matrix proved successful in providing an efficient planning framework (considering the percentages of coherent systems), the social matrix seemed to produce a less defined framework: the actors' interest in this part was more limited and the facilitators had difficulty in collecting information, which was however strategic to planning, in particular on identifying financial resources. This point therefore needs to be improved.

Finally, the simulation phase does not allow clear conclusions to be drawn on the choice of sanitation systems by the participants. More than just identifying fixed choices, this step should provide a new opportunity to contribute to the modelling process of sanitation systems adapted to the areas, which in turn would be mobilised as the process progresses.

▶▶ Sanitation planning: child's play or for experts only?

Without prior expertise on the subject, can one participate, without hesitating, in the planning of a coherent sanitation system? The results of the PLANISSIM project is a promising step in this direction!

At the crossroads of social engineering and sanitation process engineering, the project's approach involved a diverse group of actors: citizens, CSO members, technicians in the sanitation sector, humanitarian non-governmental organisations, researchers, etc. Field visits, surveys, training, construction and application of WasteWAG led to more than a thousand participations in the process. The one down side was the low participation of government services in this experimental process.

From a methodological standpoint, success is twofold. The multi-functional WasteWAG tool helped bridge previously separate methods. But above all, the main challenge was overcome in the planning phase, i.e. integrating technical knowledge into a simple model, using the same tool with several functions to remain accessible to all, proposing thematic matrices to build coherent systems.

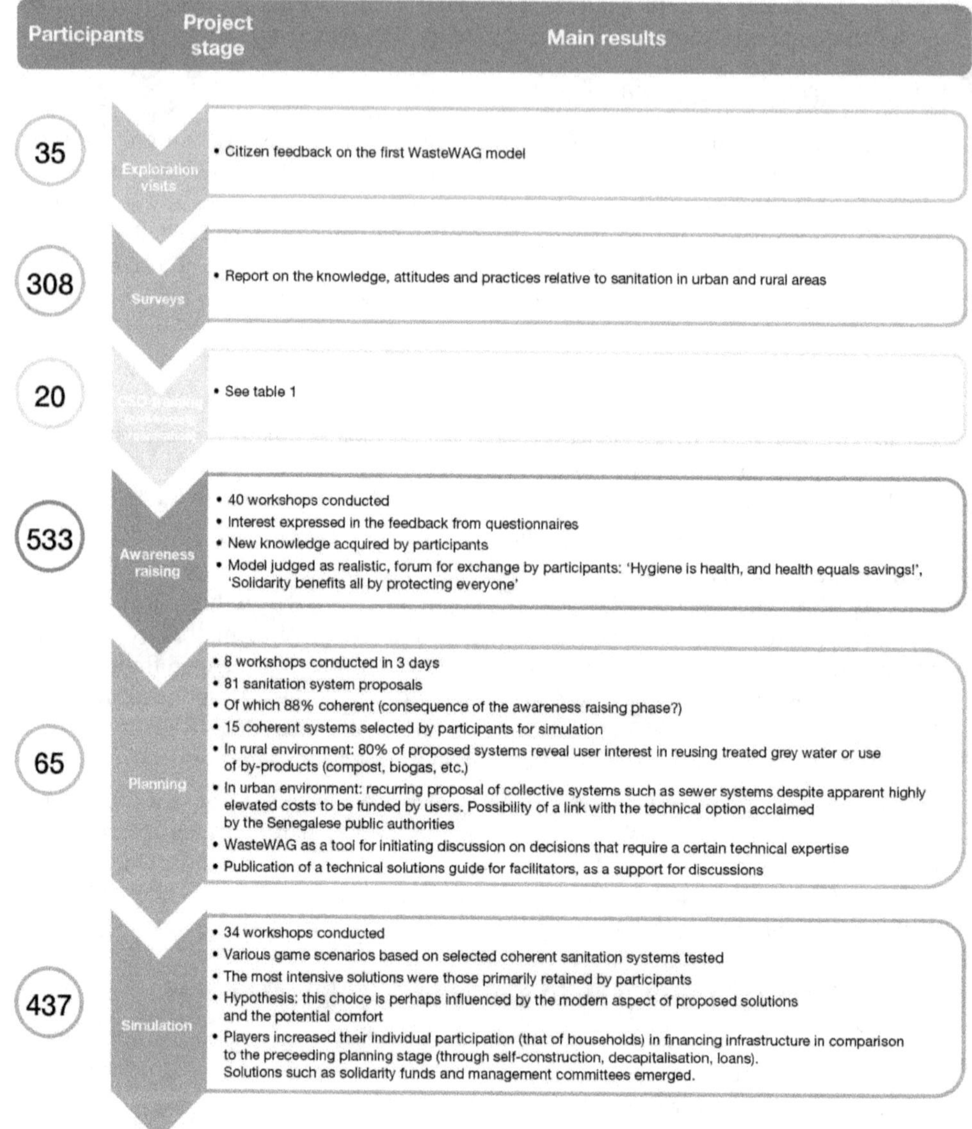

Figure 14.5. Main results of the WasteWAG application phases

Regarding the different WasteWAG functionalities over time, conclusive results were obtained in the awareness-raising and planning phases. Whereas the technical matrix proved not only to be relevant, but also to be central to proposing coherent systems, the social matrix requires further development in future work. The simulation phase revealed a weakness in the approach to choosing systems adapted to the context. Moreover, the preparation of this phase was too dependent on the intervention of the researchers; simpler methodological resources need to be found in the future to link the planning and simulation phases.

On the aspects of empowerment and transfer, the dynamic created with the CSO members was essential to the project. The sequence of training, tool construction and field application phases definitely enhanced the facilitators' skills. This provided the project with valuable modelling partners and empowered facilitators in the field, who were able to pass on technical knowledge using simple materials and guide participants in better understanding their roles in the sanitation service.

The PLANISSIM project has thus laid the foundations for a multi-actor and multi-level approach to wastewater system planning using the WasteWAG construction mechanism. This is the first step towards the approach being used in real conditions, i.e. for the process' results to be implemented. With this in mind, two projects are currently under construction: one in Guinea Bissau in an urban context and the other in Senegal in a rural environment.

The approach will also be extended to the reuse of treated wastewater by integrating a module on health risk management. Here too, projects are under way to use the approach in France, on issues that have been raised for compliance with recent European regulations.

▶▶ **References**

Arnstein S.R., 1969, A ladder of citizen participation, *Journal of the American Institute of planners*, 35(4), 216-22. https://doi.org/10.1080/01944366908977225 4

ComMod Collective, 2004. *Charte ComMod : la modélisation comme outil d'accompagnement – Posture*, Version 1.1 du 5 avril 2004. http://cormas.cirad.fr/ComMod/pdf/ComModCharte2004.pdf

Étienne M., Du Toit D., Pollard S., 2011. ARDI: a co-construction method for participatory modeling in natural resources management, *Ecology and Society*, 16(1):44. http://www.ecologyandsociety.org/vol16/iss1/art44/

Étienne M., 2010. *La modélisation d'accompagnement : une démarche participative en appui au développement durable*, Versailles, Éditions Quæ, 384 p.

Ferrand N., Abrami G., Hassenforder E., Noury B., Ducrot R., Farolfi S., *et al.*, 2016. *Coupling for Coping, CoOPLAaGE: an integrative strategy and toolbox fostering multi-level hydrosocial adaptation*, in: Proceedings of the ACEWATER2 Scientific Workshop, Accra, Ghana, 31 oct - 3 nov 2016, Ronco P., Crestaz E., Carmona Moreno C. (eds.), Ispra: European Union, 58-63. https://ec.europa.eu/jrc/en/publication/proceedings-acewater2-scientific-workshop-accra-ghana-31-oct-3-nov-2016

Gabert J., Frenoux C., Guillaume M., 2010. *Choisir des solutions techniques adaptées pour l'assainissement liquide*, pS-Eau/PDM, Guide méthodologique n° 4. http://memento-assainissement.gret.org/IMG/pdf/memento-assainissement-chap2a.pdf

Le Page C., 2017. *Simulation multi-agent interactive : engager des populations locales dans la modélisation des socio-écosystèmes pour stimuler l'apprentissage social*, Mémoire d'habilitation à diriger la recherche, Université Pierre et Marie Curie, Paris.

Tilley E., Ulrich L., Lüthi C., Reymond P., Zurbrügg C., 2014. *Compendium of Sanitation Systems and Technologies*, 2nd Revised Edition, Swiss Federal Institute of Aquatic Science and Technology (Eawag), Dübendorf, Switzerland, 176 p. https://www.eawag.ch/fileadmin/Domain1/Abteilungen/sandec/schwerpunkte/sesp/CLUES/Compendium_French/compendium_fr_2016.pdf

UNDP (United Nation Development Programme), 2015. Sustainable development Goals, https://www.undp.org/sites/g/files/zskgke326/files/publications/fr/SDGs_Booklet_Web_En.pdf

WEDC, Water, Engineering and Development Centre, 2014. *Prévention de la transmission des maladies féco-orales*, Loughborough University. https://wedc-knowledge.lboro.ac.uk/resources/booklets/G020FR-Prevention-de-la-transmission-des-maladies-booklet.pdf

Chapter 15

Influence at the margin: Participation and water infrastructure in the Cambodian Mekong delta

Jean-Philippe Venot, William's Daré, Etienne Delay, Sreytouch Heourn, Lamphin Lor, Malyne Neang, Somali Oum, Raksmey Phoeurk, Sopheaktra Say, Sophak Seng and Sreypich Sinh

This contribution reflects on an on-going participatory research process initiated six years ago in Cambodia. Taking as a starting point the duality of the literature on participation in development—as emancipatory or yet another expression of technocratic power, we explore the scope that serious games offer to understand and influence the way water infrastructure projects supporting agricultural intensification in the Cambodian Mekong delta are designed and implemented. We stress that recognising, rather than brushing aside, the fact that serious games constrain participants in different—and sometimes unexpected—ways, allows being more realistic about their effects, which we argue amount to a significant influence at the margin.

⯈ A bridge between two parallel takes on participation, research and development

In the so-called Global South, participatory research approaches raise specific issues in relation to the ways *participation* and *development* have become entangled over time. Broadly speaking, the literature on the topic can be divided into "two camps" that seldom talk to each other—on the ground that they would deal with different processes and realities. On one side, critics who point to the failure of participatory approaches writ-large to live up to their emancipatory ideal and to the fact that they have become, at best, yet another depolitising instrument in the toolbox of development agencies and, at worse, a way to deepen existing power relationships (e.g. Cooke and Kothari, 2001). On the other side, participatory research scholars, some of which focusing on participatory modelling (that can include the use of serious games or not; e.g. Voinov and Bousquet, 2010). These argue that such approaches stand in stark contrast to other more "mainstream" participatory approaches due to the fact that (1) the design of the approach itself is participatory and the (2) tools developed introduce some "distance" with the real world, which in turn acts as a buffer to limit the expression of power relationships, or at least allows unravelling them in dispassionate ways.

As a consequence, the broader critique of participation in development would not have any hold on these processes.

Establishing a dialogue between these schools of thoughts and epistemic communities, we argue, constitute a productive avenue. Their insights, when used in conjunction, can help understand what is at play in specific participatory research initiatives. The critique of participation reminds us that, in development contexts, participation (participatory research included) is generally engineered "from the outside" by development workers, researchers, or policy-makers and takes place in "invited spaces" where natural resources users are invited to contribute, but within the boundaries set by others (Cornwall, 2004). But being cognizant of these boundaries does not mean they are impermeable or that they cannot be redrawn. On the contrary, it helps identifying specific windows of opportunity (Daré and Venot, 2018) to design and implement participatory modelling processes that indeed allow exploring alternatives and can lay the basis for transformations that will do other things than just reinforcing dominant modes of knowledge and practices—and the vested interests underpinning these.

▸▸ The context: the Cambodian Mekong delta and water infrastructure projects

The Cambodian Mekong delta stands in stark contrast with its iconic neighbour in Vietnam, well known for its extended network of dikes and canals that shapes a landscape home to millions of people and intensive agricultural practices. The area remains flooded four to six months every year, supporting small scale capture fisheries and, when the flood recedes, a mosaic landscape made of a multitude of geometric fields where farmers cultivate a variety of crops (fruit trees, vegetables, rice) slowly emerges.

The area is crisscrossed by hundreds of drainage canals that also provide irrigation water for cultivation in the summer. Some of these, dating back to the early 19th century, are called *"preks"* and result from joint (1) man-made interventions in the form of breaches in the levees of the main rivers, and (2) hydrological dynamics as floods further widened the breaches and sediments deposited progressively, raising adjacent land, hence forming the long canals and landscape that can be observed today (figure 15.1). Over the last two decades, the *preks* have been seen as a means to intensify agricultural production. The Cambodian government, with support from several aid agencies, invested in their re-excavation (they had become silted-up) and in the construction of water control infrastructures (mainly sluice gates) to increase water availability in the dry season (Venot and Jensen, 2021).

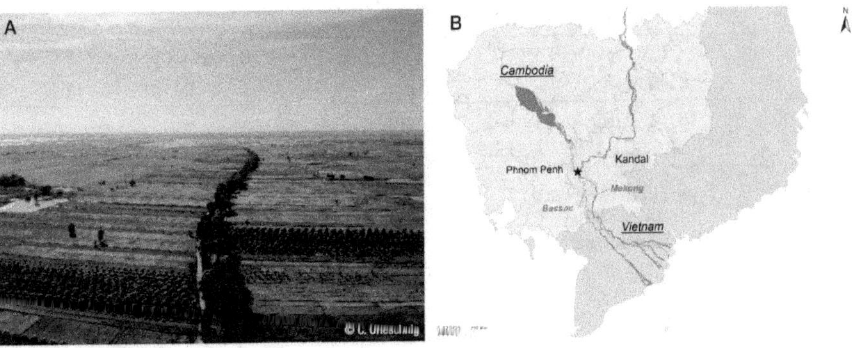

Figure 15.1. The Cambodian Mekong delta and the *prek* landscape

187

▸▸ Articulating participatory research and development projects

The initiative we reflect upon was "engineered from the outside" by foreign researchers who successfully submitted a project proposal to a European research call. The project, called DoUbT, involved a Cambodian University (the Royal University of Agriculture) and a Cambodian NGO (the Irrigation Service Centre) with whom foreign researchers had engaged with previously. It proposed to study knowledges and practices of land and water management in south-East Asian deltas, including in the Cambodian Mekong delta. One of the reasons to chose this area was that it had been less intensively studied than its Vietnamese counterpart and held the promises of generating new (academic) knowledge. Due to the interest of some members of the foreign research team, the project proposed to experiment with Companion Modelling (Etienne, 2014) and serious games as a way to generate hybrid (academic/non-academic) knowledge relating to delta management. The idea was also to train Cambodian researchers and civil society actors in the development and use of participatory modelling approaches that they could then use in other activities if they deemed these relevant.

Between 2016 and 2022, seven multi-stakeholder workshops that constituted "invited spaces" of participation were organised by the project team. These workshops brought together a diversity of actors: farmers, fishermen, village authorities, local elected officials, representatives of the decentralised government and of sectoral ministries at different levels of responsibility, as well researchers and staff from development agencies (table 15.1).

Table 15.1. The participatory process unfolded over six years and is still on-going

Date	Type of activity	Design team	Participants
April 2015	First encounter with the *preks* during study of AFD investment in the Cambodian Irrigation Sector		
March 2016	DoUbT project starts		
September 2016	J.-P. Venot based in Cambodia, in RUA		
February 2018	CIRAD exploratory visit to Cambodia		
June 2018	DoUbT Meeting IRD/CIRAD/RUA/ISC		
November 2018	Design and implementation of the first serious game	IRD, CIRAD, RUA, ISC	Participants day 1: farmers, fishermen, village chiefs Participants day 2: commune elected representatives, districts officials, ministries and development agencies staff, and researchers

Date	Type of activity	Design team	Participants
May 2019	Design and implementation of the second serious game named: *Dai Prek*	IRD, CIRAD, RUA, ISC	Participants day 1: farmers, fishermen, village chiefs Participants day 2: commune elected representatives, districts officials, ministries and development agencies staff, and researchers
September 2019	Design and implementation of a game to build the capacity of water user group in Pursat province	ISC	Participants (three days): farmers, fishermen, village chiefs and commune elected officials
November 2019	Design of a game to discuss collective action for safe agricultural production in the *preks*	IRD, CIRAD, RUA, ISC	Not implemented
December 2019	COSTEA project starts and DoUbT project finishes		
February 2020	Design and implementation of *Dai Prek*	IRD	Participants (one day): MoWRAM staff and WAT4CAM experts
June 2020	Design and implementation of a game in four *preks*	ISC	Participants (three days): farmers, fishermen, village chiefs and commune elected officials
August 2020	Design and implementation of *Dai Prek*	IRD, RUA, ISC and WAT4CAM experts	Participants (two days): farmers, fishermen, village chiefs, commune elected officials and district officials (from two different areas)
May 2022	Implementation of *Dai Prek*	IRD, ISC and WAT4CAM experts	Participants (one day): Representative of sectoral ministries and district administration, and WAT4CAM experts
July 2022	Implementation of *Dai Prek*	IRD, ISC and WAT4CAM experts	Participants (three days): Elected Representatives at commune level

AFD, Agence Française de Développement – French Development Agency; COSTEA, COmité Scientifique et Technique de l'Eau Agricole – Scientific and Technical Committee for Agricultural Water; DoUbT, Deltas' Dealings with Uncertainty project; ISC, Irrigation Service Center (Non-governmental organisation, Cambodia); MoWRAM, Ministry of Water Resources and Meteorology, Cambodia; RUA, Royal University of Agriculture, Cambodia; WAT4CAM, Water Resources Management & Agricultural Transition for Cambodia Project

To understand how the participatory process unfolded and with what effects, it is necessary to take a step back. Rather classically, research started by an exploration of the Cambodian Mekong delta. This included "field visits" along the *preks* as well as open-ended discussions with a diversity of people. Through these, researchers came to the realisation that water control infrastructures were being built in the floodplain. Indeed, at the same time than the research was conducted, a multi-million agricultural development project was implemented by the Ministry of Water Resources and Meteorology (MoWRAM) of the Royal Government of Cambodia. The project was financed through a loan of the French Agency of Development (AFD) and French (mostly engineering) experts supported their Cambodian counterparts (for more information, see Venot and Jensen, 2021).

Construction of water control infrastructures took place even though development agents and engineers had little knowledge of the extremely complex local hydrology (which was reduced to a single indicator: daily water levels of the main river). Embracing an international policy model and the Cambodian national policy of Participatory Irrigation Management and Development (PIMD), the agricultural development project aimed at enhancing the participation of farmers in the management of the rehabilitated infrastructures. Yet, the approach followed remained largely technocratic as observed in many irrigation projects that have been implemented in Cambodia and beyond over the last three decades (Fontenelle, 2020; Ivars and Venot, 2018). In practice, farmers were invited to meetings during which development agents presented decisions they had already taken on the basis of desk work and short field visits and explained the responsibilities farmers should assume once the construction works will be over. Farmers had little say on a project implemented in their name and that impacted their life and, here, the critics of participation in development have a point! This had detrimental effects, even if considered from a narrow vantage point, as some of the water infrastructures collapsed right after their construction under the effect of rather average floods.

This diagnostic determined the overall orientation of the participatory research process. Investments in water control infrastructures were likely to continue as AFD and the Royal Government of Cambodia negotiated a follow-up project; foreign researchers together with their Cambodian academic and civil society partners hence considered that it was important (1) for a diversity of people to express their views about the present and future of the *preks* and (2) to identify alternatives to the current development pathway predicted on further water control and agricultural intensification—this in line with international academic debate relating to deltas socioenvironmental vulnerability. This decision was taken in the absence of any explicit demand to do so from people living along the *preks* (though they expressed concerns about past development interventions), let alone from people involved in designing and building water infrastructures. We, the authors, set the stage and we hoped some of the knowledge generated would "seep into" the development project being negotiated.

In line with these two objectives, we developed serious games that aimed at unravelling local knowledge about the *preks* as well as inhabitants' concerns and priorities. But maybe more importantly, by using the games with institutional actors, we also aimed at questioning the idea that building water infrastructures and intensifying agriculture in the floodplains of the Cambodian Mekong delta was the "obvious" (and only) approach to follow. By confronting development agents and engineers, first-hand, with tools that they tend to frown upon and dismiss often on the ground that they are not "scientific enough", we also hoped that they start considering these as legitimate knowledge-making approaches. The idea was that recognising the relevance of the approach would, in turn, legitimise the outcomes of the sessions organised with local stakeholders and make it more likely for these to be accounted for in future development projects.

Engaging actors who were a priori unconvinced about the interest of our approach required a lot of "discussions behind the scene" and creating "interessement" (Akrich *et al.*, 1988a). Identifying individuals who were keen to experiment (whether this was because they were curious, had used similar approaches in other contexts, or just wanted to break from their routine) proved crucial. These individuals, then, acted as "spoke-persons" (Akrich *et al.*, 1988b) or "brokers" (Lewis and Mosse, 2006), helping us identifying or creating windows of opportunity for research approaches or results to be injected in the activities planned or implemented under the agricultural development project.

▶ Serious games do constrain...

By now it must have become clear that the participatory research process was neither neutral nor, some could say, very participatory given that the objectives and approach used to achieve these were defined by researchers, at least for a large part. This constituted a real boundary (in the sense of limitation) to participation; and other boundaries were also inscribed in different ways in the tools we designed (Venot *et al.*, 2022). These tools evolved over time and they also differed depending on who participated to the workshops (whether farmers and local officials or actors involved in the design or construction of water infrastructure). Broadly speaking, however, they all served as artefacts through which it was possible to describe agricultural and water management practice and discuss the expected impacts of a series of interventions (from the construction of sluice gates and roads to the organisation of farmers' training or the support of small-scale capture fisheries) on agricultural production, the environment, and ultimately local livelihoods, in a context of variable and uncertain water availability.

Participants assumed the role of farmers, local officials, or agents of sectoral ministries. They had to choose from a series of options materialised by vignettes and place these on a board that represented the *prek* landscape, thanks to a series of plywood tiles that could be assembled in any possible way either to show a familiar or a totally imagined place (figure 15.2) so as to indicate where they thought specific interventions ought to be implemented. Hydrological conditions were simulated thanks to a dice-roll and the impact of each intervention (on agricultural production, the environment, and livelihoods) shown through a pre-defined, yet explicit, calibration. Running the game several times in a row or in parallel sessions allowed the emergence and discussion of different scenarios that materialised many possible developments.

This short description (for more see Venot *et al.*, 2022) hints as yet another series of potential obstacles to the expression of participants' concerns. After all, we predefined the elements of the *prek* landscape that were represented (canals, roads, agricultural fields); its evolution was envisioned through the prism of (pre-identified) interventions that related to water and agriculture (not health, education, or rural infrastructures such as road and electricity—though these are likely to be important concerns too); parameters considered (agricultural production, environmental conditions, livelihoods) were limited and loosely defined; and impacts on these parameters were pre-calibrated. All of this stemmed from the interests and knowledge of foreign researchers who initiated the research activities and the scope of on-going development projects they aimed to inform and influence.

Figure 15.2. A version of the serious game used with local stakeholders in June 2020

Yet, these constraints did not really hold in the face of practice. One of the central tenets of Companion Modelling is that participants can modify the tools proposed—redesigning them in the process for future iterative use- and this is exactly what happened. Participants re-shuffled the tiles of the board as they deemed fit, they identified other types of intervention than those proposed, calculations were made with little regard to the calibration (and sometimes not made at all), and participants interacted with each other as they wanted, regardless of any instruction we may have tried to enforce. This is not mere tinkering; participants played an active role in re-shaping the serious games.

▸▸ ... But they also influence at the margin

That serious games are oriented towards specific objectives (set by those who design them), and that they constrain participants accordingly, do not mean that they can only serve dominant powers. The participatory research process we initiated had transformative effects, at least at three levels.

First, engineers whose job is to design water control infrastructures, and who were at first reluctant (to say the least) to recognise that local farmers might have something to say about these, started recognising the value of local knowledge. This was illustrated, for instance, when one of these engineers multiplied one-to-one discussions asking participants clarifications about what they had done or said during game sessions and carefully noting down the information. Second, staff of the Cambodian NGO who had contributed to designing the game sessions fully reinvented these when they used the plywood tiles in activities they conducted to support the establishment of Water

User Associations in another province of Cambodia. This demonstrates the malleability of the tools used but also their ability to make sense of the approach in their own terms and transforming it in the process. Third, *preks* are now envisioned in a very different light than they had been in recent development projects. The studies that underpinned the participatory research process stressed the largely ad hoc nature of engineering interventions that treated *preks* as if they were independent, almost disjointed, water channels. The serious games, on the other hand, stressed their interconnected nature and articulated a vision of the *prek*s as one among many elements of a mosaic landscape (represented by the plywood tiles in figure 15.2). This latter vision materialised in the sense that the development project that is now being implemented (and that we aimed at influencing) does not use single *preks* as its scale of intervention but, rather, what development agents call "*prek* development areas", that is, groups of adjacent *preks* that are hydraulically connected. Further, the scope and duration of feasibility studies have been extended and (some) development agents seem keen to continue experimenting with "active" participatory approaches that go beyond mere consultations. Such changes remain fragile as they also go hand-in-hand with delaying infrastructure works to the dismay of other actors.

Our activities resulted in subtle yet tangible changes in terms of how participation of farmers is envisioned and the scale at which *prek* rehabilitation is planned. The overall doctrine—that development in the Cambodian Upper Mekong delta hinges on building water infrastructures and intensifying agriculture—has not changed however. This is understandable. After all, this is what engineers—who continue to steer most irrigation development projects—know and do. This is also a very visible way to demonstrate that "something is happening", which is a prime concern of decision makers. The participatory research process fell short of one of its key ambition, to articulate strikingly different trajectories to the current development path. Rather, the effects we highlight are modest changes, which is why we talk of *influence at the margin*, yet they are important. The development projects that are underway provide an opportunity to see whether these modest changes can lay the basis for more significant transformations, and notably whether planning at the landscape level translates into practices that give more room to *prek* users and the environment.

▸▸ Conclusion

It will not come as a surprise to most readers that participatory research, and more specifically participatory modelling, is not a "miracle solution". We hope, however, to have shown that engaging with the critique of participation in development can be a useful way to reflect on participatory research.

▸▸ References

Akrich M., Callon M., Latour B., 1988a. A quoi tient le succès des innovations ? 1 : L'art de l'intéressement. Gérer et comprendre. *Annales des Mines*, 11: 4-17.

Akrich M., Callon M., Latour B., 1988b. A quoi tient le succès des innovations? 2 : Le choix des porte-parole. Gérer et comprendre. *Annales des Mines*, 12: 14-29.

Cooke B., Kothari U., 2001. *Participation: The New Tyranny?* London, Zed.

Cornwall A., 2004. Introduction: New democratic spaces? The politics and dynamics of institutionalised participation. *IDS Bulletin*, 35(2): 1-10.

Daré W., Venot J.P., 2018. Room for Manoeuvre: Users participation in water resources management in Burkina Faso. *Development Policy review*, 36: 175-189.

Etienne M. (Ed.), 2014. *Companion Modelling: A Participatory Approach to Support Sustainable Development*. Dordrecht, Springer.

Ivars B., Venot. J.P., 2018. Entre politiques publiques et matérialité : Associations d'usagers et infra-structures d'irrigation au Cambodge. *Nature, Sciences, Sociétés.* 26 (4) : 383-394.

Fontenelle J.P., 2020. Délégation de la gestion des systèmes d'irrigation aux irrigants : peut-on éviter de reproduire les échecs du passé ? *In* S. Bouarfa, F. Brelle, C. Coulon (Eds). *Quelles agricultures irriguées demain ? Répondre aux enjeux de la sécurité alimentaire et du développement durable.* Versailles, Quae, p. 137-154.

Lewis D., Mosse D., 2006. *Development brokers and translators: The ethnography of aid and agencies.* Bloomfield, New Jersey, Kumarian Press.

Venot J.P., Jensen C.B., Delay E., Daré W., 2022. Mosaic glimpses: Serious games, generous constraints, and sustainable futures in Kandal, Cambodia. *World Development,* 151 (2022):105779.

Venot J.P., Jensen C.B., 2021. A multiplicity of prek(s): Enacting a socionatural mosaic in the Cambodian upper Mekong delta. *Environment and Planning E: Nature and Space,* 5(3): 1446-1465.

Voinov A., Bousquet F., 2010. Modeling with stakeholders. *Environmental Modelling & Software,* 25(11): 1268-1281.

Chapter 16

Enhancing the characterisation of floods, water resources and their impacts through participatory observation

Valérie Borrell Estupina, Frédéric Grelot, Alexandre Alix, David Badoga, Pierre Balzergue, Pauline Brémond, Moustapha Djangue, Jofre Herrero Ferran, Alain Fezeu, Camille Jourdan, Cécile Llovel, Linda Luquot, Valérianne Marry, Sylvie Morardet, Roger Moussa, Diana Puigserver Cuerda, Sophie Richard, Marine Rousseau, David Sebag, Eric Servat and Sandra Van-Exter

Water shortages, violent floods, intense run-off: in the face of global change, regional trajectories are evolving, at times deviating from those of the past. Reliable information is needed to renew knowledge and define prevention and remediation policies. How can the transfer of data from the field be diversified and multiplied? Participatory observation processes no longer uniquely lie with professionals but involve ordinary citizens as well, and thus can contribute to this goal.

▸▸ Challenges of participatory observatories in relation to socio-hydrosystems

Changes and socio-hydrosystems

Climate forces and anthropic pressures exerted on the environment induce changes which are the source of ruptures, thresholds and non-stationarity in the behaviour and functioning of hydrosystems, modification of their balances, and the emergence of new risks (French National Institute for Earth Sciences and Astronomy, INSU[1]). The impacts on populations are manifest at several levels: destruction of their property, in their activities, in their use of the resource, and even their lives. Populations thus need to adapt if they do not want these same crisis situations to be repeated.

These hydrological and societal changes are non-stationary; it can be expected that future operating conditions of these systems will significantly deviate from those in which current knowledge has been established. This makes it difficult and uncertain to extrapolate the past and present functioning and evolution of socio-hydrosystems for use in future scenarios.

---

1. Institut national des sciences de l'univers: https://www.insu.cnrs.fr/fr/los-services-nationaux-dobservation

The need for observations that bear witness to these changes has thus increased. The log data usually collected for hydrometry (rainfall, river levels, piezometric levels) must be reinforced as much as possible, and log data obtained through observation of the impacts on the sectors at stake and of the behaviour of populations must be enhanced. Diversifying the nature of these observations would also be beneficial in obtaining a more holistic vision of a watershed, where the processes, particularly physical, biological, social and political (internal and external) interact to lead the socio-hydrosystem towards its future trajectory.

Long- and short-term observatories

Reliable information is needed to renew knowledge and define prevention and remediation policies. To this end, observatories are deployed at different levels (national long-term observation systems from the INSU; regional water observatories operated by public stakeholders to share knowledge and help with decision-making; ephemeral observatories run at the local level by motivated groups). The following paragraphs pinpoint areas for improvement.

When a data log exists for a site, the observed data may contain spatial or temporal gaps and uncertainties. Thanks to advances in forecasting for watersheds not measured by the international hydrological scientific community (Hrachowitz *et al.*, 2013), even short-term observation records can now be used in hydrological modelling with acceptable uncertainties (Jourdan, 2019).

The nature of the data collected can also be extended to less conventional hydrological data. Testimony from local residents on the state of their crops or river flooding over the course of the seasons provides information on the evolution of rainfall and run-off, for example. The data collected must however be validated before it can be integrated into scientific processes.

The changes that socio-hydrosystems undergo can also be analysed through the impacts that they cause, such as damage (household, infrastructure, ecosystem, pollution, health, etc.) and various adaptation dynamics. Participatory observation thus becomes multidisciplinary, providing a broader and more exhaustive understanding of the consequences of events and their evolution.

Hence, an increase in heterogeneous observations in space and over time is of great help to scientists and decision-makers. So how can this data from the field be multiplied and diversified? This is where the interest of participatory observation becomes clear.

Participatory observatories

Participatory observatories are a common and unifying term for a family of polymorphous devices that allow an audience to participate in the collection or production of knowledge with epistemological or socio-political implications. Participatory hydrology can be seen at least as "a method of collecting hydrological data [...] that allows each citizen to contribute to the improvement of scientific knowledge" (Hassenforder, 2020). It can also cover the entire process of co-production of hydrological knowledge between stakeholders and researchers. Participatory hydrology can then be used to respond to scientific questions as well as societal and environmental issues (Hassenforder, 2020). Participatory science can refer to "participation in science by amateur audiences to extend scientific knowledge", or "the participation of science in citizen concerns" (Mitroi and Deroubaix, 2018). Their evolution over the last century is presented in figure 16.1. Originally highly focused on the production of biodiversity inventories, they have been implemented over the years in a wider range of domains and have consequently given rise to participatory observatories in the 2010s and the production of hybrid knowledge. However, the scientific validity of the data produced should continue to be scrutinised when producing this knowledge.

The reader may refer to Mitroi and Deroubaix (2018) for the results of their almost exhaustive literature review that identifies participatory science initiatives in the field of water. These initiatives are still experimental and geographically diverse. Public participation is most often integrated through the use of inexpensive and easy-to-handle sensors. These sensors provide quantities which, although measured at a single point, are integral to the functioning of an entire system (such as the flow at a river outlet). The information collected is thus valuable because it is global and pluralistic.

Three examples of recently implemented participatory observatories are presented below. They present a sample of the diversity of participatory observatories and emphasise the key roles of stakeholders. They are presented in order of duration of the scheme, from the shortest (one to two weeks) to the longest (several years).

▶▶ Various examples of participatory observatories in hydrology from short to long term

An interdisciplinary and international field school in the Mediterranean region as a one-off participatory observatory?

Fifty-two percent of the world's population will be under water stress by 2050. Climate change, population growth in conurbations, tourism, agricultural practices and water resource management methods are contributing factors to this forecast. Consequently, the sustainable management of catchment areas is a crucial societal and environmental issue in the changing context of the 21st century, which basin managers must address.

But how can a basin be managed if its functioning and evolution are not sufficiently known? Meteorological observation systems are very well deployed in France, yet hydrometric observation systems remain more limited and a good number of basins do not have sufficient spatial and temporal coverage of hydrological indicators of their state to guide the trajectory of sustainable management. How can university instruction help basin managers to fill this gap in hydrometric data on specific territories? How can it promote opportunities for experts and resource management stakeholders to meet? Finally, how can it contribute to increasing interdisciplinary analysis of future evolution scenarios? Interdisciplinary field schools are a pedagogical mechanism driven by universities; they can help turn these instances into reality.

In February 2019, an international, interdisciplinary field school was set up by MUSE K-IM WATERS (University of Montpellier, AgroParisTech, SupAgro, CNRS and IRD) and the University of Barcelona, Spain, on the Muga river basin, upstream from Empuriabrava. This basin is subject to strong anthropic, tourist and agricultural pressures. A naturalised protected area has been put in place to treat wastewater and improve biodiversity. A dam was built upstream to support irrigation, drinking water and hydroelectricity needs, as well as to control floods, but it is no longer sufficient to satisfy the uses.

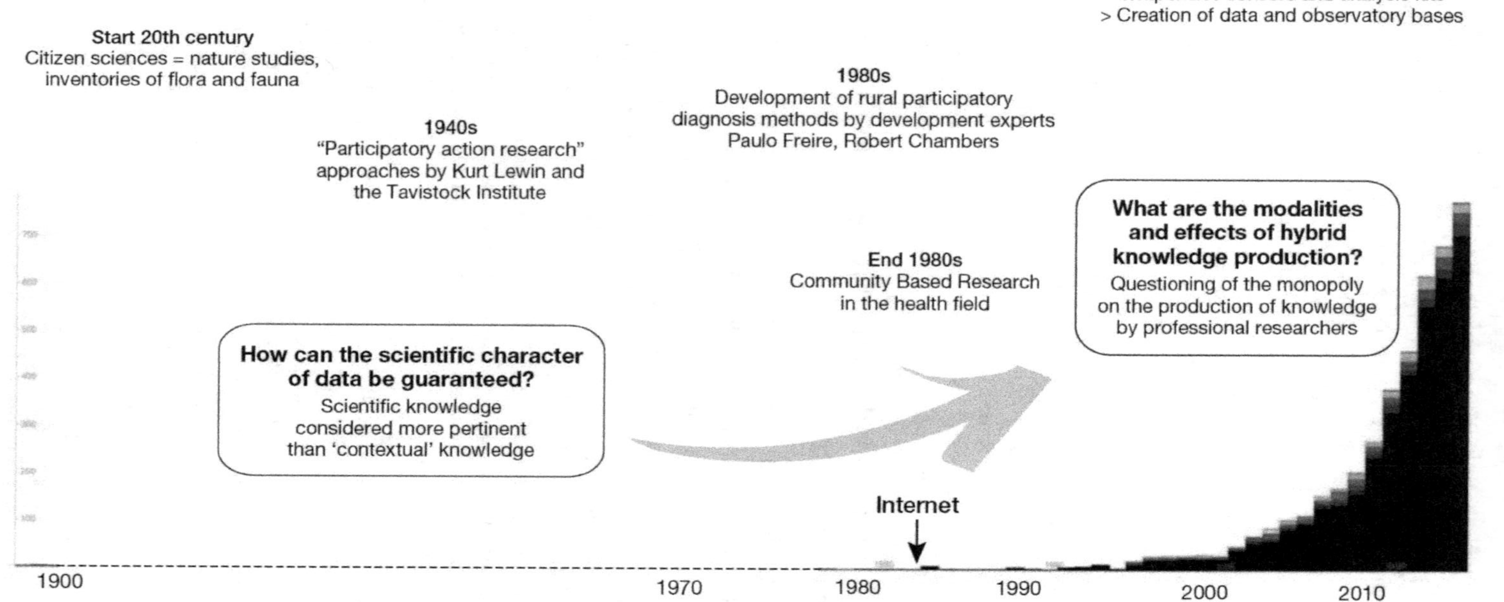

Figure 16.1. Evolution of the number of publications in participatory science (inspired from Mitroi and Deroubaix, 2018; Houllier and Merilhou-Goudard, 2016, modified; Hassenforder, 2020)

Involvement of several audiences, each with distinct roles, is essential to the smooth running of the interdisciplinary field school and to achieve the participatory objectives:
– The teacher-researchers, who oversee the action, ensure the inclusion and participation of everyone throughout the project, along with experts and engineers from private and public companies. They also guarantee the reliability of the protocols deployed and the scientific quality of the information collected.
– The stakeholders and managers of the catchment area bring to the group: field knowledge, events effecting the field, and management skills.
– The students are the key actors of the project bringing different cultures and backgrounds:
– They are trained by the experts and by their peers in the technical or scientific aspects;
– They collect data (measurements, analyses, observations, testimonies) within the scope of a specific protocol;
– They enrich the feedback from various experiences on this site (conferences and visits with stakeholders and experts) through direct questioning and confront the actions carried out with the issues at hand;
– Lastly, they may even set up future management scenarios for debate to evaluate different adaptation opportunities (role-playing games, workshops, discussions).
– Residents, who are asked about their experiences and changes in their practices following recent floods or water shortages, can also contribute to the process through their participation in surveys, the results of which are structured and analysed. In short:
– hydrometric observations and data of a diverse nature are collected, evaluated and structured;
– the interactions between physical, biological and social processes are better understood;
– factors threatening the long-term management of the basin's water resources are identified.

The academic participants recognise that the participatory project is above all pedagogical: the students are there to learn and the teachers to teach... The active pedagogy, used to solve a concrete problem in a real setting (field, actors, issues) turns the students into actors, observers and bearers of new views on the issues and problems of the basin. Exchanges and debates with local stakeholders help develop a more detailed and global understanding of the issues.

A holistic understanding of the issues and problems requires not a multi- but a transdisciplinary approach. The shared points of view, consultation and participation of the various stakeholders are essential to achieving this objective. The participatory nature of the field school is a way of bringing together visions, analyses and actions, opening the way to interdisciplinarity. The learning and implementation of transdisciplinarity would however require more specific prior training in these approaches.

A low-cost participatory hydro-meteorological observatory in a tropical climate

The population projection for Africa is 2.5 billion people by 2050, of which about 55% will live in metropolitan areas. The population of Yaoundé, for example, has grown from 90,000 in 1960 to over 3.65 million in 2017. This enormous increase has led to significant urban and agricultural expansion, to the detriment of forests and wetlands. Coupled with the effects of climate change, these changes in population density and land use have a considerable impact on socio-hydrosystems (Jourdan, 2019).

Understanding and reducing the impacts of these changes on run-off, flooding and pollution requires the observation of indicators of the state of these socio-hydrosystems. The need for observation is all the more urgent as, in the past, local hydro-meteorological monitoring services have not always been able to carry out their observations due to economic, health or civil crises.

The doctoral work by Jourdan (2019) thus proposed to develop a low-cost hydro-meteorological observatory for the Méfou catchment area, including the metro area of Yaoundé, over a short period of two years (March 2017 to March 2019) by involving the local population (experts, technicians, amateur citizens). The objective was comprehensive and consisted in:
– ensuring hydrological and rainfall monitoring using low to moderate cost field equipment to produce quality data for scientific use;
– collecting feedback on the overflow levels, well filling levels and geo-referenced run-off zones affected by heavy rainfall in order to diversify the modes of validation, even partial, of the operating hypotheses;
– training master students, doctoral students and technicians in scientific techniques, protocols and issues;
– raising awareness of the local population on the issues involved in this research and the risks to which it is exposed.

This work was made possible thanks to close collaboration between scientific institutions from the North and South (University of Montpellier, INRAE, CNRS, IRD, Universities of Yaoundé I and Ngaoundéré, Centre de Recherche en Hydrologie du Cameroun, Philiae Ingénierie). The participation of several groups of locals with distinct roles was essential to the success of the experiment:
– The doctoral student from the north acted as an expert and pilot for the operation;
– A local technical expert was specifically trained by him for the duration of the project to ensure a field relay, and to allow an initial evaluation of the quality of the data collected in situ (an essential process prior to the scientific validation of the data) (figure 16.2);
– The handlers or one-off observers (master students, doctoral students and local technicians) assisted the local technical expert in carrying out scheduled data collection and one-off experiments (figure 16.2);
– The primary role of the amateur citizens (local residents and farmers) was to ensure the protection of installed equipment against acts of vandalism. In this way, they served as a relay to raise awareness among the general public. They could also take photos during rainfall events, collect information on overflow levels, and activated run-off areas. These collections did not require the handling of instruments, which are the source of great uncertainties, or even incorrect measurements, when not handled correctly or not used in suitable environments.

Participants in the field (pilot, expert, handlers) were acknowledged in the form of scientific and technical learning. The technical expert notably benefited from particularly advanced training that was recognised by the Cameroonian hydrological agencies.

Figure 16.2. Local expert collecting data from a limnimetric station in Yaoundé in 2016

The amateur citizens, especially the local residents, were appreciated for their responsibility for and protection of the equipment (sometimes installed on their property) and were acknowledged by the general public as knowledgable on these issues.

In this example, which led to the publication of scientific reference works, the trio of scientist, expert, and amateur ensured dynamic commitment to the project over the two-year period and that the objectives were reached.

A flood impact observation system (so-ii) in the Mediterranean climate

The impacts of floods are often reduced to the negative consequences of these phenomena in terms of damage, expressed in financial terms. This perspective is largely influenced by systems that bear the consequences, whether they be insurance policies or special public compensation programmes. In this scope, impacts are measured by the compensation received by "disaster victims", which is supposed to allow for the replacement or repair of damaged goods. This, however, does not take into account all the consequences that do not give rise to financial compensation, such as:
– the consequences of weak events for which compensation systems are not mobilised;
– consequences that appear later than the time of compensation, such as premature wear and tear of material goods;
– certain non-material consequences such as psychological impacts.

This scope also fails to take into account the dynamic aspects of the consequences of flooding: how long did it take for repairs to take place? Nor does it consider all the strategies that the affected people may have put in place following the flood to respond to it, either in the short term, during the crisis itself, or over the long term as part of an adaptation strategy.

In order to have the necessary means to observe these impacts in their entirety, the "so-ii" project set up a network of impact observers who include individuals, entrepreneurs and farmers. The objective of this network is to document in a detailed and sustainable manner the consequences suffered by these people, the possible links between material impacts, impacts on activity or use, and the impacts specific to people, as well as the strategies that they themselves may have put in place. This project was also an opportunity for the various observers to network with each other in order to share their experience, as well as to share with the academic world through a critical interpretation of the observations collected.

Starting in early 2020, the network was built up through a combination of strategies:
– "passive" communication was ensured via a website dedicated to the project, relayed by social network tools mobilised by the syndicates in charge of flood management in the region;

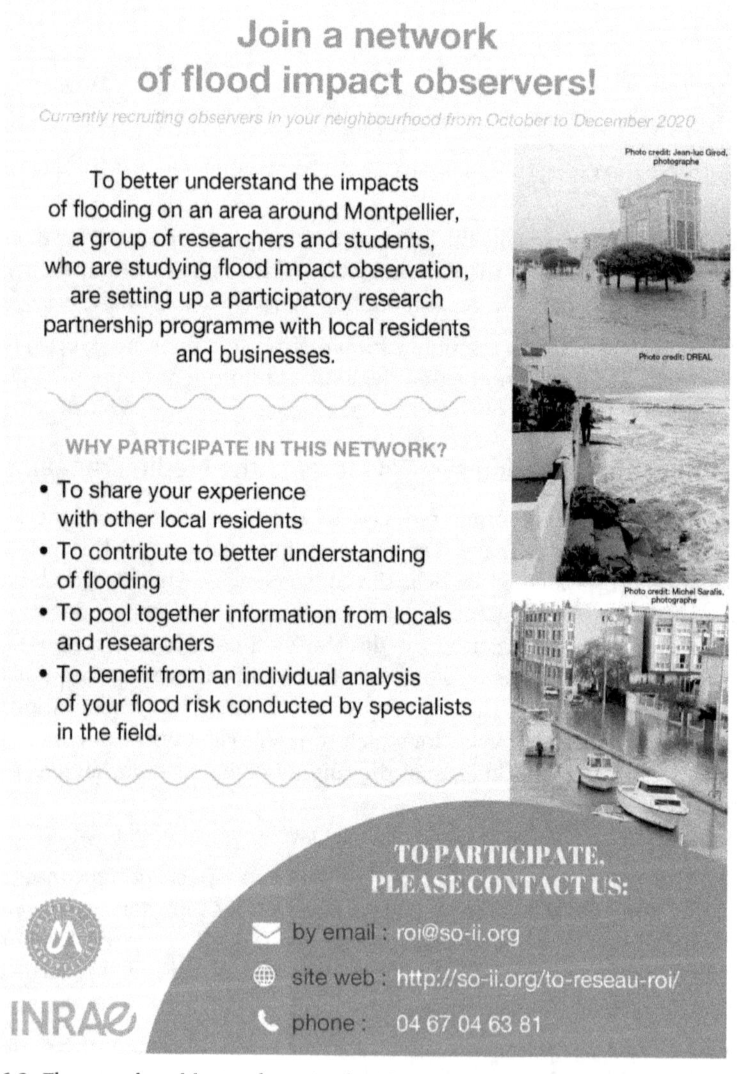

Figure 16.3. Flyer produced by students in the Master 2 programme on water-society in the scope of their 2020 interdisciplinary project (photo credit: © so-ii; https://so-ii.org/)

– targeted surveys were carried out among residents of areas known to be flood-prone (figure 16.3), with a focus on people in buildings with evidence of adaptations made after past flooding (such as cofferdams or their rails) and those that can be recommended by others (neighbourhood or association networks).

Once a person was contacted, an initial assessment was carried out together to determine whether there was mutual interest in participating in the process. If this was the case, the person was integrated into the network to become an impact observer. A more in-depth diagnostic provided a detailed picture of the person's exposure, particularly through the "modelling" of their building. This stage provided initial feedback to the observers, and also allowed for the person's use of the building to be monitored, as well as to monitor in detail the consequences of flooding. Currently, sixteen households are candidates for the scheme. A more in-depth diagnostic is underway in their homes.

From 2021 onwards, observers are contacted each year to take stock of events related to the subject at hand:
– Have they experienced flooding? If yes, consequences are carefully documented, even if this entails contacting each person several times.
– Have they carried out or modified any of their projects in relation to their place of residence or activity? If so, the exposure assessment is updated.
– Finally, each year, a meeting is organised to allow exchange between all observers, flood managers and researchers involved in the process. The first meeting with the network candidates was planned for December 2020.

The direct expectations of the scheme are twofold:
– to improve knowledge of the impacts of flooding, by ensuring that this knowledge is shared with those primarily concerned: the local populations in a region, the managers of the phenomenon.
– to organise direct sharing of experiences, both positive and negative, between people living in the same area.

Mainly developed for private individuals for the time being, the aim is to implement this network of observers in the agricultural and business sectors in the near future.

▸▸ References

Hassenforder E., 2020. Place de l'hydrologie participative dans les recherches avec, par, sur ou pour la participation multi-acteurs. Conférence présentée au 19e séminaire Eau organisé par Polytech Montpellier le 26 février 2020.

Houllier F., Merilhou-Goudard J.B., 2016 Les sciences participatives en France : Etats des lieux, bonnes pratiques et recommandations, 63 p. https://hal.science/hal-02801940/

Hrachowitz M., Savenije H.H.G., Blöschl G., McDonnell J.J., Sivapalan M., Pomeroy J.W., *et al.*, 2013. A decade of Predictions in Ungauged Basins (PUB) – a review, *Hydrological Sciences Journal*, 58 (6), 1198-1255. https://doi.org/10.1080/02626667.2013.803183

Jourdan C., 2019. *Approche mixte instrumentation-modélisation hydrologique multi-échelle d'un bassin tropical peu jaugé soumis à des changements d'occupation des sols : cas du bassin de la Méfou (Yaoundé, Cameroun)*, Doctorat de l'Université de Montpellier, École doctorale GAIA, OSU OREME, HSM, LISAH, Soutenue le 05-12-2019 à Montpellier. http://www.theses.fr/2019MONTG089

Mitroi V., Deroubaix J.-F., 2018. « Faire sciences participatives » dans le domaine de l'eau. Trajectoires croisées au Nord et au Sud, *Participations*, 21(2), 87-116. https://doi.org/10.3917/parti.021.0087

Planning for Transforming towards Socio-Ecological Sustainability

Chapter 17

CoOPLAN multi-scale participatory planning process: Applications in Uganda and elsewhere

Nils Ferrand, Clovis Kabaseke, Moses Muhumuza, Thaddeo Tibasiima and Emeline Hassenforder

This chapter introduces CoOPLAN, a specific approach for participatory planning aiming at enabling a group of participants to co-construct together a collective action plan to change together in their environment. The chapter provides a detailed description of the various steps of the CoOPLAN process and illustrates how it was implemented in a specific case in Uganda. The chapter also includes a comparative discussion of the implementation of the CoOPLAN approach in four cases (Uganda, metropolitan France, New-Caledonia, Tunisia). It highlights the modifications that were made to adapt the approach to the specific context of each case.

Participatory planning, as the design of an action plan by a group, is the essence of strategic decision-making for governments, business or any community. It aims, initially, at anticipating and organising a complex set of actions, responding to stakeholders' needs and coping with an uncertain environment. It also structures stakeholders' commitments, identifies ways to share resources, builds a vision of a common evolution and hence strengthens social links. The planning process should obviously produce a plan; but it is a key social learning process. It has its own transformative value (Smith, 1973) by engaging participants in sharing and aligning their expectations, their options' proposals, and their understanding of the future. It helps discussing on resources, dependencies, commitments, risks, solutions and may thereby set conditions for a more resilient and adaptable society. Planning and adapting become complementary: the future adaptation processes are themselves planned, by including a monitoring and steering apparatus.

In this chapter, we introduce CoOPLAN, a specific approach for participatory planning extended from participatory modelling. CoOPLAN has been developed by researchers from the G-EAU joint research unit "Water Matters" in Montpellier, France, and extended internationally since 2006. The second part of the chapter details how the CoOPLAN approach was implemented in a specific case in Uganda. The third part of the chapter presents a cross-reading of four CoOPLAN processes implemented in different contexts: Uganda, France (mainland), New-Caledonia and Tunisia. The chapter compares the four processes and highlights the adaptations they led to.

⟶ Components and steps of the CoOPLAN process

The implementation of a CoOPLAN process includes various components, steps, actors and tools, which we will present here. As stated in the introduction, the overall aim of a CoOPLAN process is not only to produce a plan, but also to strengthen the social ties between participants and thus create favorable social conditions for implementing the plan. Formally, CoOPLAN is a participatory modelling process, which uses two meta-models[1]:

– one to let participants propose actions, through an "action sheet", pre-instantiated based on a common action meta-model for all CoOPLAN process, and later instantiated by the participants in many "action proposals";

– a second one to let them structure action proposals in plans, through an "integration matrix", which follows itself a meta-model, and is pre-instantiated in a specific matrix for this application case.

Figure 17.1 summarises the components of the CoOPLAN process. The process includes one or more stakeholder groups who will co-design one or more plans. This is specified below. Stakeholders include a "pilot" who is the participatory process initiator and leader, and a "pilot group" which gathers supporters of the pilot.

The pilot, and eventually other stakeholders, start by organising the process (see step 1 in table 17.1). This implies selecting and engaging participants, preparing logistics and materials and communicating about the process (see chapter 9 for more details). The first step in building the plan is for participants to identify common objectives, stakes or goals (step 2). It is also during this step that they define the spatial, temporal, thematic and scale boundaries of their future plan. For this, they can draw on existing diagnostics, if any. The participants then propose various proposals of actions to achieve these objectives (step 3). The resulting list of action proposals is shared with all participants. These action proposals are then sorted into thematic categories (e.g. agriculture, health, education, etc.) (step 4), and discussed and detailed through the filling of "action sheets". An action sheet is a material instance of an action "meta-model", i.e. structured components allowing to build and use a model for a given purpose. The action sheet is one of the central components of the CoOPLAN approach (along with the integration matrix presented below). The same meta-model is kept throughout the whole process. The action sheet allows to specify (1) the resources needed to carry out the various action proposals, with an estimated intensity (e.g from 0 to +3) of this requirement, (2) the expected impacts of these actions, also with an intensity (positive or negative, or both), (3) the scales at which the actions are to be implemented, and (4) where and when the action is to be carried out (figure 17.2). The elements of this action sheet, i.e. the choice of resources, impacts and scales mentioned in it, can be made by the participants themselves or in advance by the pilot, the pilot group and/or experts (step 5). These "action sheets" are then completed individually or in small groups (step 6), followed by a comparative dialogue during which participants compare the various action proposals with each other and improve or modify the content of the related action sheets (step 7). This database of action proposals can then

---

1. A meta-model, in this chapter, is a set of types of concepts and rules, or grammar, which allow to build a given type of model. In practice, it can be a language or a method. Providing a meta-model to modelers steer them toward a given family of models, for some repeatable purpose.

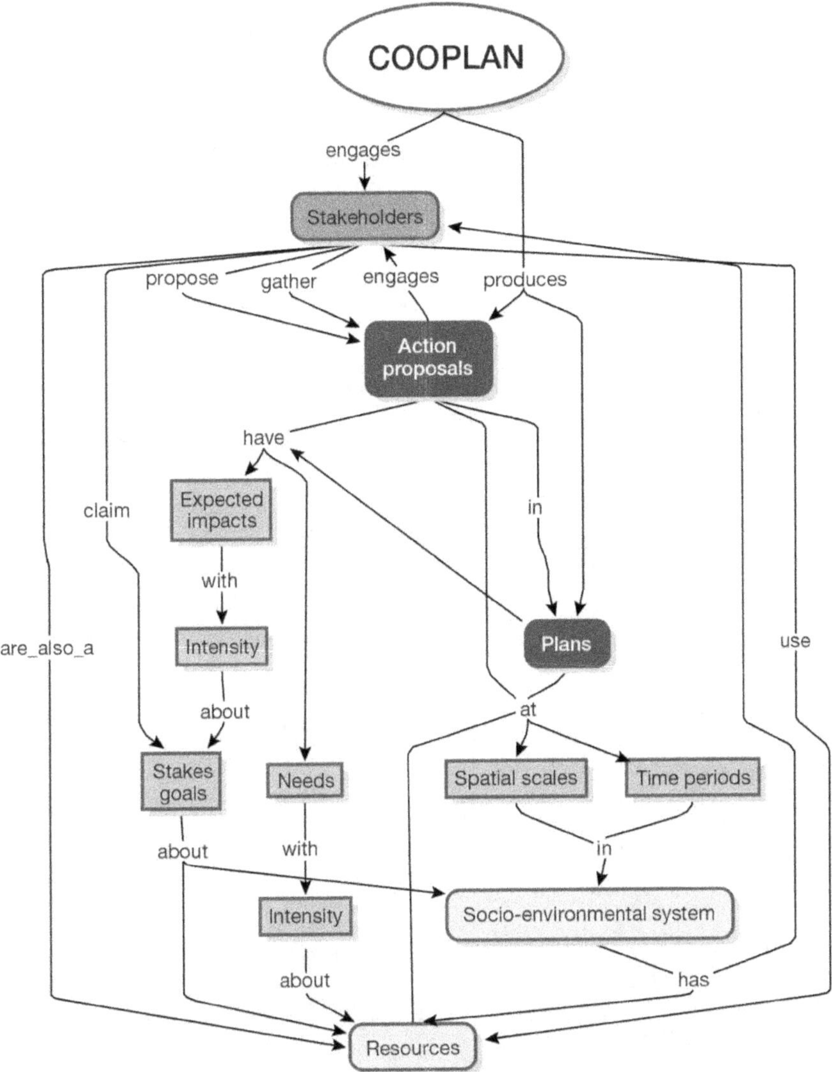

Figure 17.1. Components of the CoOPLAN process

be analysed (by the pilot, experts and/or participants) to check the consistency of the actions between them (step 8). For example, participants will check that action A does not require a larger budget than action B, whereas the participants had evaluated them on an equal budget basis.

The group then moves on to constructing the plan itself. To do this, participants select the action proposals that they feel are most relevant to achieving the desired objectives, and structure them in a logical, temporal and spatial manner in a first version of the plan (step 9). This structuring is based on CoOPLAN's second central component, which is also a meta-model: the integration matrix. The integration matrix is an empty grid that incorporates the various components of the action sheet: resources,

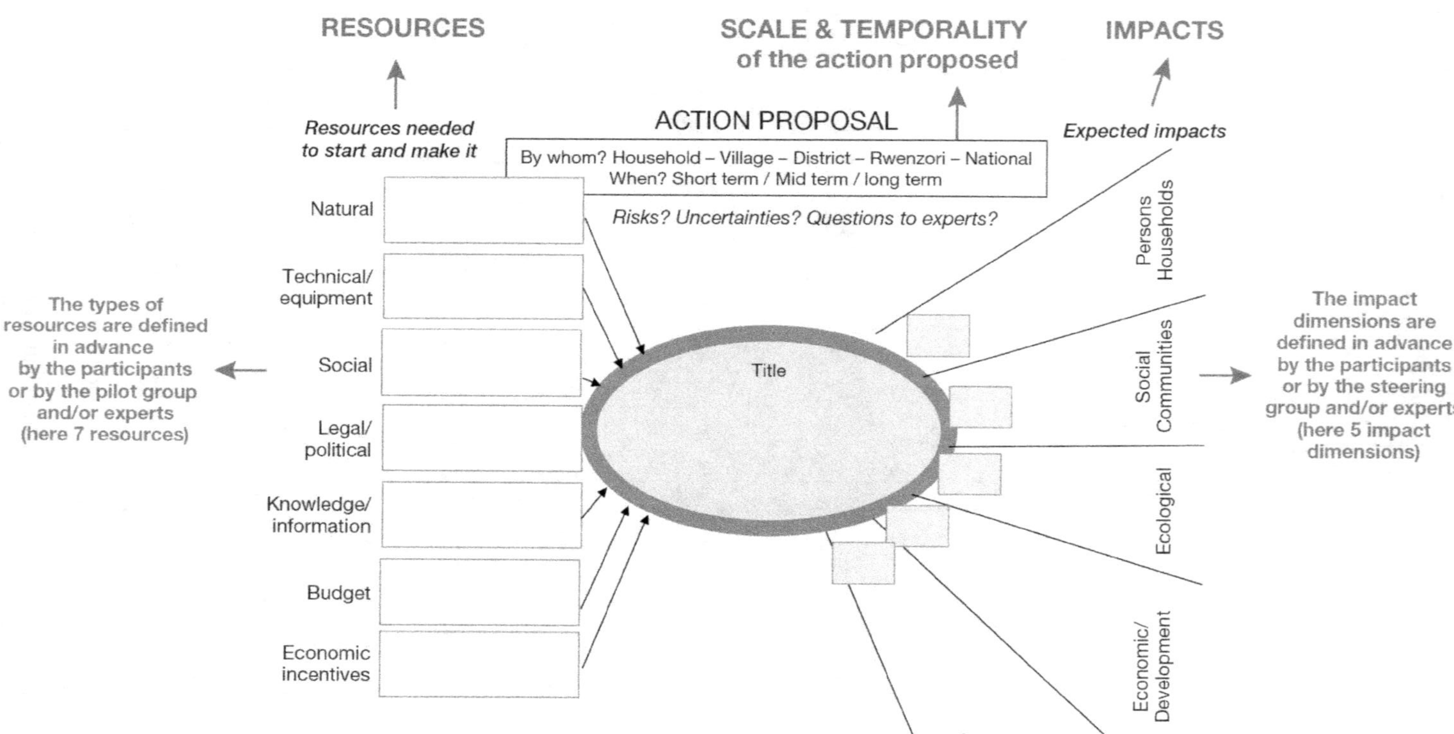

Figure 17.2. Example of action sheet and its components (resources, scale and temporality, impacts)

impacts, scale and temporality of action implementation (short-term, mid-term, long-term, see figure 17.3). The matrix is accompanied by a spatial map enabling the actions to be precisely located when relevant. Stakeholders position the action proposals selected in the matrix according to the timeframe and scale at which they are to be implemented. Then, for each target scale, on the basis of the information entered in the action sheets for the selected proposals, they assess the intensity of the global resources' requirements, column after column. In the same way, they assess the impact of the various actions on the different impact dimensions defined beforehand. The impacts of the actions can be positive or negative, or kept positive and negative if different arguments are combined. For example, the creation of a hillside lake can increase available water resources, but it can also destroy vegetation, create siltation or restrict access to water for users who do not have access to the lake. Looking at the matrix, for each target scale, participants then assess the feasibility (resources requirements) and efficiency (achieving the expected impacts) of their plan (step 10). For instance, if they look at the natural resources column, and see that many actions selected will require a lot of water or land, they must discuss and adapt the actions, withdraw some actions or add additional ones (e.g compensation, provision) to make sure that the plan is feasible and efficient.

Once they have obtained a plan that seems feasible and efficient, participants can test this plan in a participatory simulation (e.g. role-playing game with WAG) or by simulating the impact of extreme or "stress-test" scenarios and discussing their effect (e.g. what if a flood or a migratory wave occurs in the area?). They can adapt the plan accordingly. If several plans were produced by different groups, participants can then integrate and choose one final unified plan by comparing globally all plan alternatives, hybridising among plans or choosing one (step 12). They then need to re-assess the final plan as in step 10. The final step involves communicating on the final plan and formalising a commitment act from participants to symbolise their future involvement in the implementation of their plan (through a signature or else).

Most of the steps outlined above can be carried out in face-to-face workshops or online. However, given that the process aims to strengthen the social ties between participants, it seems necessary for certain key steps to be carried out face-to-face (notably steps 7, 9, 10, 11 and 12). Similarly, each of these steps can be more or less participatory, i.e. carried out by the pilot alone, by all participants and/or involving other stakeholders (pilot group, experts, etc.). These choices are made when engineering the CoOPLAN process (see chapter 9).

In summary, CoOPLAN is a deliberative, integrative and structured plan design mechanism with contradictory evaluation. CoOPLAN does not replace multiple technical-scientific expertise, it is complementary. It is not a multi-criteria method of group decision-making, but an assisted social protocol for discussing collective action. There is no aggregative or arithmetic process for exhibiting a better solution. Participants have to gather proposals and discuss them with the support of their own products. It is impossible, in principle, if the process is truly participatory, to know or impose in advance the diverse dimensions or sectors of the future actions, as these could have a wide scope linked to the participant's visions. Thus, the method is totally open in regards to the scope of proposals made, which can be technical, social, organisational, etc. Nevertheless, some steps indubitably constrain

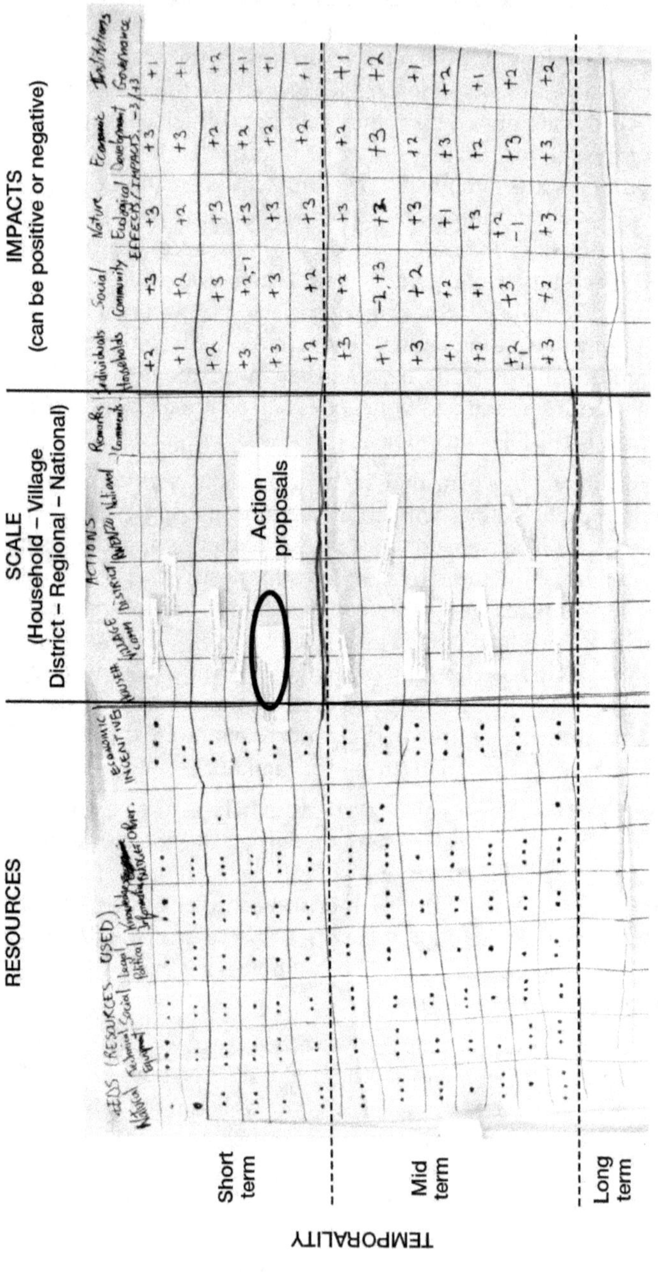

Figure 17.3. Example of integration matrix and its components (resources, impacts, scale, temporality)

the "spectrum of possible" that emerge in the process, notably step 1 (selection of participants and modes of engagement), step 2 (framing spatial, temporal, thematic, and scale boundaries of the future plan), step 4 (clustering of actions) and step 5 (elaboration of the meta model). To reduce the biases induced by these steps, it is possible to carry them out in a participatory way, but this can also increase the duration and cost of the process as a whole.

We have provided here a detailed description of the various steps of the CoOPLAN process (table 17.1), which can give the impression of a long and complex process. Nevertheless, several of these steps can be carried out in the same workshop, as we illustrate in the following section with the example of Uganda. Furthermore, if the process is to strengthen social ties between participants, it needs to be spread over a period of time, but not over several years, otherwise it risks creating participation fatigue. A more detailed description of the CoOPLAN process in Uganda can be found in Hassenforder (2015).

Table 17.1. Detailed steps of the CoOPLAN process

#	Step	Goals	Actors
1	**Organisation**	Organise the conditions of the process	Pilot
2	**Normative framing**	Define boundaries and objectives, what the plan aims at changing	Pilot + Pilot Group or all participant (if inclusive participation)
3	**Action proposals**	Get participants to propose ideas of actions	Pilot + ALL
4	**Actions' synthesis**	Organise action proposals in thematic clusters	Pilot + experts
5	**Common framework for describing action proposals**	Select relevant scales, resources and impacts for describing action proposals (i.e. define the content of the action sheets)	Pilot + pilot group + experts
6	**Detailed actions' description**	Fill the action sheet for each action proposal	Pilot + pilot group or ALL
7	**Comparative dialog**	Share and improve action sheets	Pilot + ALL + experts
8	**Consistency/ Harmonisation**	Global comparison of action proposals to check consistency	Pilot + ALL + experts
9	**Plans' design**	Select and gather actions in a logical, temporal, and spatial manner to build an action plan	Pilot + ALL
10	**Plans' assessment**	Analysis of the plan to assess feasibility and efficiency	Pilot + ALL
11	**Testing plans**	Test plans by simulation or robustness analysis	Pilot + ALL
12	**Plan selection**	Integrate and choose one final unified plan	Pilot + ALL + experts
13	**Finalisation**	Officialise the final plan	ALL

▸▸ Case: planning for integrated natural resource management in Uganda

In the European project Afromaison[2] (2011-2014), researchers used CoOPLAN to support integrated natural resource management in five study areas, including in the Rwenzori Mountain range in Uganda. The Rwenzori region is located in western Uganda, at the border with the Democratic Republic of Congo (Figure 17.4). It covers 14,000 km^2 with a population of about 2,4 million. This region of mountain tropical forests has several environmental assets, with fertile soils. Predominantly inhabited by smallholder farmers engaged in subsistence farming, it also hosts some commercial farming, and a significant touristic activity.

Inhabitants' subsistence practices such as bush burning, fuel wood harvesting and unsustainable timber harvesting have led to deforestation, soil and ecosystems degradations (Plumptre, 2002). Combined with climate change and high population growth rates, it led to food shortages and disease outbreaks (Migongo-Bake and Catactutan, 2012). This makes the region economically vulnerable with a majority of people below the poverty threshold (Ubos and ILRI, 2007).

Uganda has a fairly comprehensive list of natural resource management legislation and policies. From 1992, natural resource management was devolved to the local governments (Onyach-Olaa, 2003), shaped by a five-tier structure (district/county/subcounty/parish/village). Environment committees and officers are responsible for community engagement and implementation of natural resource management laws. However, lack of governmental funds, heavy workloads and corruption impede adequate implementation of this legal framework. Other important issues include problems of land tenure due to the reinstatement of the former traditional kingdoms in 1993. Few international donors are still active in the region. Since 2003, regional civil society organisations, later joined by other stakeholders, have gathered under a coalition called the Rwenzori Regional Development Framework (RRDF, 2011).

This CoOPLAN process was initiated, piloted and facilitated by six local researchers (also farmers themselves) from Mountains of the Moon community University (MMU) in Fort Portal, supported by French and Belgium researchers of the Afro-Maison project. Local facilitators originate from the area, belong to the cultural and linguistic groups and are involved in natural resource management through a pre-existing network of community organisations (with farmers field school and other training or sensitisation activities). Five "rapporteurs" were also hired to monitor the process in the communities.

The European researchers proposed a set of initial methodological trainings for the Ugandan partners; which allowed them to implement CoOPLAN at both the regional and the local level, with a joint dialogue. The aim was to support regional and local stakeholders in the co-construction of a multi-scale natural resource management plan. European partners have supported the Ugandan partners with a very restricted direct intervention with participants.

---

2. AfroMaison website: http://www.afromaison.net (consulted April 10, 2015).

Opportunities and challenges of transferring the CoOPLAGE approach to Integrated Water Resources Management in a Beninese wetland

Raphaëlle Ducrot and William's Daré

This chapter presents the added value and the challenges of a CoOPLAGE planning approach to operationalise Integrated Water Resources Management in 22 villages of the lower Ouemé valley, Benin. After a capacity building phase in 13 pilot villages, facilitators replicated the approach on their own in a second set of nine villages. Participants appreciated the engaging capacity building process and the mobilisation power of the role-playing game used to support the choice of actions. Although the objective was to support institutional bricolage mechanisms, in practice the project framework constrained the engagement in socio-political issues questioning the long-term outcomes beyond action implementation.

Operationalising and sustaining Integrated Water Resource Management (IWRM[1]) policies in deltas require facilitating the participation of local populations in the identification and implementation of actions to be undertaken, thereby ensuring the mobilisation and commitment of local populations over time. Ad hoc structures (such as management committees) are often set up to ensure this participation as well as to serve as an interface between the local population and water management projects. Large-scale engagement of the community and good governance are often viewed as a key requirement to build the legitimacy of such structures. Aware of these challenges, an NGO asked for support in developing an approach to facilitate the mobilisation of stakeholders in the implementation of development activities within an IWRM development project intervening in the lower Valley of the Ouémé delta in Benin.

Two Cirad researchers (an agro-geographer and a sociologist) offered to provide a support going beyond the development of communication and mobilisation strategy to engage the stakeholders in a participatory planning process, inspired by a CoOPLAGE[2] approach (chapter 2), while also factoring in equity issues. The objective

---

1. Gestion Intégrée des Ressources en Eau (GIRE)
2. CoOPLAGE (Coupler des Outils Ouverts et Participatifs pour Laisser les Acteurs s'adapter pour la Gestion de l'Environnement) = Coupling Open and Participatory Tools to Let Actors Adapt for Environmental Management.

was also to create conditions that support the development of institutional "bricolage", i.e. conditions adapted to the local context (Merrey and Cook, 2012; Cleaver, 2012; Booth, 2012; Batchelor *et al.*, 2000). It aims at allowing for individual or collective experiences to be shared and local institutional innovations perceived as fair and legitimate based on local social networks, to emerge. In other words, rather than tools and methods, what matters is the posture and process.

The NGO agreed with this principle and acknowledged the need to offer space for the expression of a variety of viewpoints to be expressed, notably through the engagement of institutional actors and local stakeholders, to remain flexible in the implementation process and encourage institutional adaptations. Yet, our analysis of the participatory process shows that the frame of the project proved unadapted to ensure that the necessary specific posture was transferred rather than the tool.

This chapter looks back at the hiatus that finally emerged between the application of the principles of the participatory approach and the project constraints to which the NGO was subjected. The aim is to gain a clearer understanding of the determinants of this hiatus and to draw useful lessons for the future IWRM development projects.

After this introduction, the next section elicits the NGO request. The third section introduces the way we addressed this request, and show the different steps of the participatory process. The fourth section discusses the lessons learned and recommendations.

▸▸ The request: engaging stakeholders to support the creation of water committee

The project was part of a Dutch cooperation program in the Beninese water sector, called OmiDelta[3], more specifically funded by the Non-State Actors Fund (ANE) managed by SNV[4]. The ANE launched two calls in 2018 to fund project for operationalising IWRM in the lower and middle Ouemé Valley, notably in the Beninese coastal delta. The NVW-GIRE (Nouvelle Vallée de l'Ouémé GIRE) project was funded by one of this call more specifically targeting IWRM and aimed at promoting the valorisation of water resources, resilience to flood and drought, prevention of erosion process and improvement of the water resources governance in the basin. Domestic water supply and sanitation activities were funded under a second call and food security activities were excluded from funding. The propositions were also to have explicit consideration of good governance, innovation, gender, youth mobilisation and climate changes issues. In total six projects were funded under the ANE program which established a unique quantitative Monitoring and Evaluation (M&E) system to monitor the progress and outcomes of the 13 projects with periodical meetings to discuss methodological issues and results[5].

---

3. OMIDelta Fond Acteur Non Étatique, SNV, s.d. Un instrument de financement des ANE actifs dans AEPHA et GIRE. https://a.storyblok.com/f/191310/accec64dd5/plaquette_snv_omidelta_fonds_ane.pdf
4. "SNV is a mission-driven global development partner working across Africa and Asia. Our mission is to strengthen capacities and catalyse partnerships that transform the agri-food, energy, and water systems, which enable sustainable and more equitable lives for all" (www.snv.org).
5. OMIDelta Fond Acteur Non Etatique, SNV, s.d. Services AEPHA améliorés et GIRE locale opérationnalisée grâce à 13 projets. https://a.storyblok.com/f/191310/210e1e0423/2022omideltaane_livretr-c3-a9capitulatif des13projets_snv_vf.pdf

In the lower Ouémé basin, IWRM issues are related to the evolution of the socio-ecological functioning inundation plains affected by increases in demographic, land and anthropic pressures over the last few decades. The waters now have high bacteriological, organic or heavy metal loads due to a very low level of sanitation as well as polluting agricultural and domestic practices (such as dumping of solid and/or liquid waste). Riverbank deforestation and certain fishing techniques based on the accumulation of branches (*Acadja*) are responsible for the gradual filling of Lake Nokoué in the south of the delta and the depletion of fish. Sand extraction is also increasing because of the area's rapid urbanisation. The dynamics of land tenure, soil fertility, as well as terrestrial and aquatic biodiversity directly or indirectly influence all of the livelihood strategies of local populations. Fishing is the main economic activity on account of the great ecological wealth of these environments. It is completed with hunting, gathering and craft activities. But agricultural activities play also an important role in the local livelihoods, as the lower delta soils benefits from the annual July to October flood. The associated flood agriculture is part of dynamic horticulture value chains which supply the main urban centers of Benin and neighbouring Nigeria, both easily accessed by river.

But the hydrological functioning of the delta is increasingly disturbed by climate change, which affects the flooding season including longer-lasting flood, and increases in saline levels coming from the mouth of the river. The locals complain about the consequences of the hydrological changes on long-cycle crops.

The NVW GIRE project mobilised the conceptual framework of ecosystem services (ES) to develop IWRM activities. ES are the benefits that people derive from ecosystems (Millennium Ecosystem Assessment, 2005)—such as the transport of goods and people through canals or the increase in fertility through the sediments of a flood. It was assumed that using this conceptual framework would facilitate the perceptions of benefits of the proposed activities within four types of ES for a river (provisioning, regulating, supporting, and cultural ES). In practice, the project aimed to facilitate the development of a local economy that values certain ES, and to highlight the dependence of development on ES (e.g. river transport disrupted by water hyacinth development). The specific objectives were to disseminate knowledge on these ES, to promote the implementation of sustainable economic alternatives favourable to ecosystems, notably the development of a hyacinth value chain around one innovative firm. Although the proposal aimed to address ES in general, the proposal targeted more specifically the river transport, erosion control, not excluding other activities.

The development of local water committees, which could later be integrated into basin water organisations that Dutch cooperation was committed to develop, ensured the sustainability of the actions undertaken by the project and their institutional anchoring. The project committed to create two types of committees, one at the local level and the other at a more regional level, the two structures of which were not predefined. The project was coordinated by the NGO Protos whose partners were a private company in charge of developing the hyacinth value chains and a consulting firm in charge of characterising ES. The NGO was in charge of proposing a methodology to identify activities, establishing the committees and subcontracting local NGOs to provide environmental mediators (EM) to implement activities in the communities. The objectives of the project, formalised in a contract between SNV and the Protos NGO, was to target 100,000 people in 22 villages in 36 months.

The project was thus looking for an approach that could facilitate the mobilisation of local stakeholders, use the ES conceptual framework to identify activities, and be easily replicated. A local consultant coordinator trained in companion modelling (Barreteau *et al.*, 2003) proposed to develop a role-playing game to disseminate the ES concept and mobilise local actors. The Consulting firm in charge of ES studies in the project contacted CIRAD for supporting the development of the methodology.

The NGO also had its own agenda and constraints in the process. First of all, as a newly intervening NGO in the IWRM sector and in the region, it wanted to establish its credibility and to build its legitimacy at the local level, in a context where the population was tired of interventions with no concrete impacts. They were also engaged to the program with specific quantitative objectives. Thus, the NGO wanted to engage in concrete actions as soon as possible. Due to the size of the target population, the NGO was looking for an approach that could be easily upscaled to 22 villages and easily transferable to local mediators. Thirdly, the NGO wanted also to address gender issues, in order that the development plans also combine views of the most vulnerable people (notably women and youth).

▶▶ The proposition: a participatory planning process

Rather than mobilising the role-playing game as a tool to disseminate the concepts of ES, we proposed to integrate it within a process that could facilitate (1) the hybridisation of ES concept with local knowledge and know-how concerning the functioning of the wetlands area and (2) the emergence of local institutions adapted to local context to ensure the sustainability of actions funded by the project, that is supporting "institutional bricolage" rather than implementing ad hoc governance bodies disconnected from local socio-political functioning.

But we identified various challenges to the approach: (1) We were not completely convinced that the ES framework was adapted to engage stakeholders into the collective mechanisms underpinned by many of the activities or the governance rules necessary to improve the access to the related services that could be proposed by the actors during the participatory process. (2) The timeframe seemed to us too tight to be able to grasp the complexity of socio-political relationships that are needed to account for long term changes to occur. Especially as sustainability of activities and their institutionalisation supposed to engage non-village stakeholders such as communes and/or other active NGOs in the planning and implementation processes. (3) Local mediators had not only to master the different tools of the CoOPLAGE[6] approach but also to develop mediating skills as they were crucial for the facilitation process to achieve the outcomes planned. Therefore, the six months schedule devoted to the planning and preparation phase was very short to develop the approach and related tools, enrol the different stakeholders in the process and build the capacities of the agents.

This context led us to propose a CoOPLAGE planning process to be facilitated by EM at village level with four main adaptations:
– the participation basis in each step of the process should vary with steps restrained to some "representatives", steps open to all and specific steps to coordinate with non-village stakeholders;

---

6. http://www.g-eau.fr/index.php/fr/productions/methodes-et-outils/item/888-l-approche-cooplage

– the intervention should be developed in two phases: one devoted to the training, development and testing of the approach in a first set of villages that would take place during the first six months of the project and the second phase for its replication in the other sets of villages;
– the role-playing game should focus on the ecological functioning of the area sustaining ES and its contribution to family livelihood but should also offer opportunities to discuss the role of socio-political links in their access;
– an explicit attention should be given to social justice and equity in the approach.

The different steps of the planning process approach

The approach included several stages (figure 19.1):
– a rapid diagnostic of the villages was undertaken to understand the village territory and the village influential people;
– a collective discussion was held on the environmental issues related to the socio-ecological functioning of the delta with the support of a role-playing game;
– proposals of actions were formulated;
– a community-actions plan based on one of the tools proposed in the CoOPLAGE kit, was constructed and validated by the villagers (called CoOPLAN);
– a discussion of these plans with non-villagers (technical services of the communes, and other institutions) in order to propose implementation plans where the role of the different actors was specified.

Each of these stages involved varying levels of participation. Restrained 'Participation' was based on the mobilisation of influential people—that is village representatives with influence capacity over other actors—and we proposed to differentiate people around gender groups (women, men, youth): during the diagnostic, EM were asked to map the village main institutions and related actors (administration, committees and organisations including religious, cultural or economic oriented one), to tell the main historic steps of village development and impacts on its territory and to identify key environmental and water issues by engaging with villagers and key actors in an informal manner and/or small group discussions. The expected outcomes were to identify key water related preoccupations of the community as well as identify village people interested in the environmental and/or water related issues linked to village development and livelihood with local "influence". In practice this would include a diversified group including elected members of the village councils, association members and some individuals with specific influence (religious...). They were to form the representative group mobilised in the following steps of the process: the identification of different possible actions, their prioritisation and the characterisation of the resource needs[7] for each action. Each person consulted in these first steps was then invited to collectively select the final five to seven actions to be prioritised on the basis of the hierarchy proposed previously in each group. Small groups (again based on gender division) drew up an action plan for each action, after which a final collective group discussion was conducted that aimed at arbitrating and ensuring consistency. This final action plan was then presented to all the villagers for discussion and validation. Lastly, the elaboration of the implementation plans themselves mobilised representatives from the village and the commune as well as external actors (other NGO...) working in the village.

---

7. Three linked resources that are money, materials, labour, as well as knowledge, rights and legal resources, and capacity for collective mobilisation

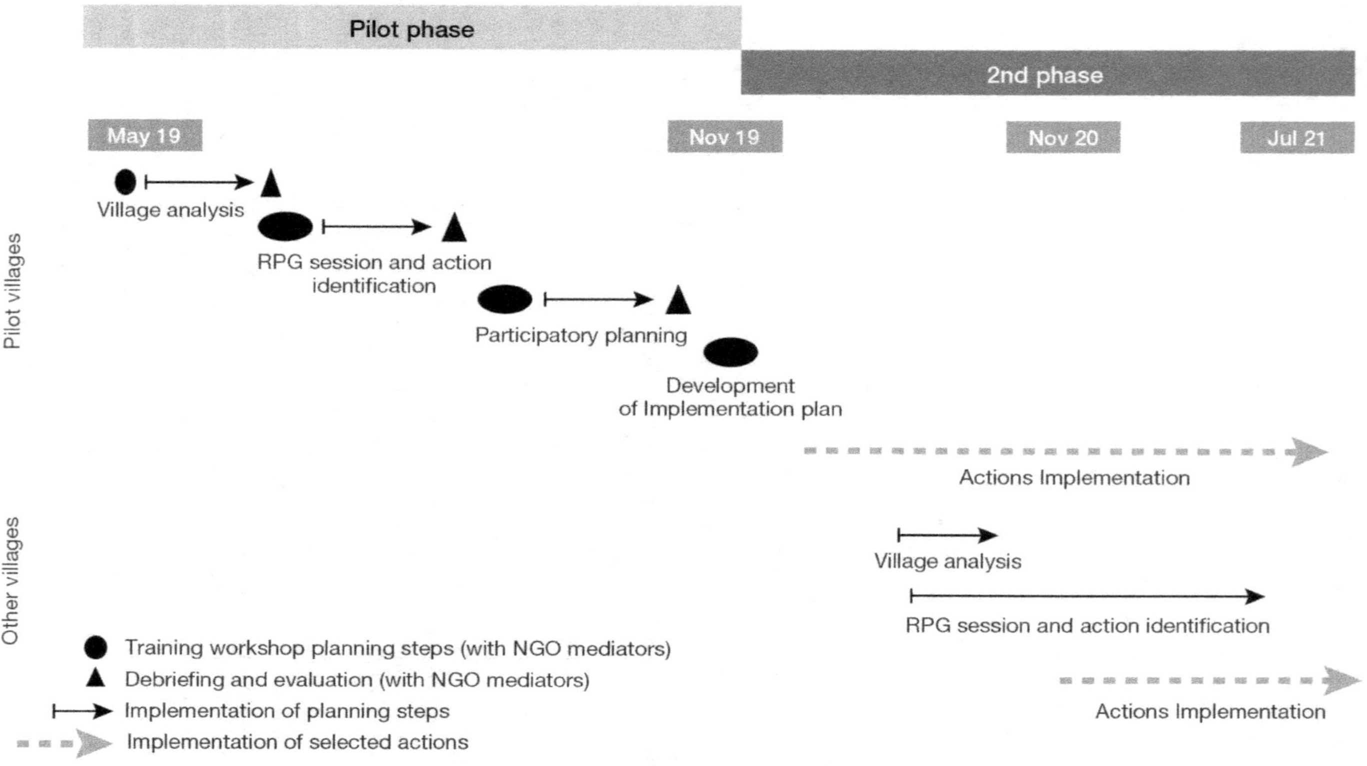

Figure 19.1. Summary diagram of the different stages of the process

A two phases process

The participatory planning process was developed in two phases. The researchers were deeply engaged in the first phase through several field trip missions. Whereas, in the second phase, due to the combination of planning constraints, researchers' interventions not funded and COVID 19 period, the NGO and its local agents had to develop the activities on their own.

The objective of the first phase was to finalise the proposed tools and train the facilitators in the process. It was conducted using a first pilot batch of 13 villages located in four target communes of the project between June and November 2019. During this phase, the two researchers, helped by a local consultant, were in charge of designing the approach, building tools and creating training for the EM who will be deployed in 13 villages, where they were expected to stay. To achieve this in the six months planning schedule, each of the work stages was preceded by a training workshop gathering EM and the process team (table 19.1). The training aimed to introduce the proposed tools, the principles of facilitation and the skills needed to facilitate the related discussions. Between each training period, the EM implemented the proposed tools in their assigned villages.

Table 19.1. The different stages of the approach and training in the pilot phase

Step	Outcomes	Date	Activities	location
1	Visit report	24-28 / 05/19	Selecting villages (NGO leadership)	villages
2	Workshop report	25/06/2019	First contact with selected pilot villages (NGO leadership)	villages
		27/06/2019	Training NGO facilitators on participatory village mapping and introducing them to role playing games	Hotel Dangbo
3	Game prototype	Jul-19	Building the first version of game prototype based on literature available on internet	France
4	Workshop report	02-06/07/2019	Workshop for testing and finalising the RPG; training NGO facilitators in the game	Hotel Dangbo
		07/08/2019	Testing the game in two villages	Villages
		08-09/08/2019	Game fine tuning and training workshop for NGO facilitators on game facilitation	Hotel Dangbo
5	13 game sessions	12-24 /08/2019	Building 13 game supports	Cotonou
6	Visit report	27-30 /08/2019	Introduction of the game in villages, on-site coaching of facilitators by NGO	villages
7	Village Game report	Sep-19	Game sessions in 13 villages with three target groups (men, women and young people)	villages
8	Workshop report	20-27/09/2019	Training workshop on the development of participatory actions plans; training facilitators on facilitation processes	Hotel Dangbo
9	Village planning report	Oct - Nov 2019	Participatory village planning in 13 villages	villages
10	Workshop report	19-23/09/2019	Training workshop on plan implementation and institutionalisation	Hotel Dangbo

The programme of the following training workshops began with a collective assessment of activities previously implemented to allow, if needed, for them to be adapted, followed by training on the tools to be used in the next phase (table 19.2). At the end of each stage of the pilot phase, the EM agents were required to provide a report on the implementation of activities per village, summarising the level of participation, the main results and main difficulties.

At the end of the first six-month pilot phase, the EM were responsible for independently implementing the seven stages of the intervention in the other nine villages, under the direct responsibility of the NGO. At the same time, they were developing the priority activities in villages of the pilot phase. In practice the second phase was initiated in the first trimester of 2020, that is at the start of the COVID19 epidemic.

An ecosystem services game

Although the conceptual model underpinning this non computerised role-playing game relies on the ES framework, the term is not used during the game session or the concept formally explained. To reduce conception time, it was chosen to model game format on a pre-existing WAG[8] that engaged into the ES service framework[9]. The game objective is to instigate discussion of the relationships between floodplain ecosystems in the delta and family livelihoods. It thus connects livelihood activities with eutrophication process, hyacinth growth, fish dynamics and availability of family labour due to health issues. ES are mobilised as qualitative indicators of the outcomes of game rounds. We chose to focus on four types of ES using around six aspects: the productive function of ecosystems (production ES), navigability (support ES), maintenance of biodiversity, environmental pollution (regulation ES), tourist attractiveness and social cohesion (cultural ES). The first five services are directly linked to different livelihood activities. The two last indicators were included to bring into the discussion emergent properties of individual actions as well as links with governance.

These indicators are being qualified collectively at the end of each round by the players. In the debriefing phase, the discussion focuses on the evolution of these indicators, how this could be connected to players' interactions and how it connects with real life situations. After this discussion, players are encouraged to identify possible actions to address the same real-life issues.

Due to the short period of time available for the building of the game, the conceptual model was based on the initial project and program reports and a literature review concerning geographical, sociological, anthropological and economic aspects of the delta. This basis was later fine-tuned with the outcomes of the village's diagnostic and tests in two villages. The game development and testing phase was used to strengthen the facilitators' capacities on the concept of ES, on developing/strengthening their facilitation skill concerning the game itself and on strengthening the facilitator's posture. The speed with which the players from the communities appropriated the

---

8. The water game (WAG) MyRiverKit was created in the scope of the European Interreg SPARE (Strategic Planning for Alpine River Ecosystems) project (https://spare.boku.ac.at/index.php/en/myriverkit).
9. The NGO has developed a short video presenting the game and its role (https://www.youtube.com/watch?v=I-g-sOWP7h8).

Table 19.2. Facilitator training workshop content in the pilot phase

Date	Type of workshop	objectives
27/06/2019	Training of facilitators	Introduction to non-computerised role-playing games with the MyRiverKit game that mobilises the concept of ecosystems services
		Participatory mapping exercises (village and stakeholder mapping)
02/08/2019	Debriefing of villages exercises	Sharing information gathered in each village and participatory comparative analysis
		Identification of commonalities and differences in the 13 villages
05/08/2019	Game workshop	Test of game prototype V1 (laboratory game)
		Introduction to facilitators of game mechanisms, general framework, the different roles and processes included, and artefacts mobilised
		Introduction to game implementation and facilitation
07/08/2019	Game session in village	Test of game prototype V2 in a real context (two villages)
		Coaching facilitators in real life situation of game facilitation (two in charge of game facilitation, two facilitators for game monitoring in two villages)
09/08/2019	Training workshop	Test of prototype V3 with NGO members (two game sessions facilitated by NGO run in parallel)
		Two facilitators in charge of each game
		Different specific exercises in the afternoon concerning the facilitation stance (statement reformulation, neutrality etc.)
19/09/2019	Training workshops	Debriefing of information collected during the game sessions with facilitators, categorisation of actions, and training of facilitators on the tool "action sheet"
20 & 21 and 23 & 24 09/19	Workshop on village participatory planning in small groups	Training facilitators in the use of different tools for building the action plans with village representatives (one session with two men/village; one session with two women/village + village chief) in the form of coaching in real planning situation
		Training facilitators on presenting the plan, qualifying actions, building action plans and putting outcome of planning exercise up for discussion
27/09/2019	Workshop with facilitators and stakeholders	Introduction to the overall approach to build the action plan with focus on how different steps relate to each other
		In a mock exercise, facilitators confronted with other facilitators on methodology steps not yet developed in the workshop (qualification of needed resources for each action) and coaching on facilitation of each steps
		Questions concerning overall approach
	Workshop with stakeholders' representatives	Coaching facilitators in real life situation to build implementation plan with representatives of villages and communes, four sub-groups each dealing with one type of action

game elements, the strategies retained and discussed, and the comments made during the debriefing confirmed that the game's economic focus and the way it was represented made sense to the community.

In practice, two different game boards were designed in order to consider the specific territorial, landscape and ecological realities south of the delta (Aguégués commune) on the one hand, and the three other communes highlighted by the initial diagnostic on the other (Dangbo, Bounou, Adjohoun). The principles and calibration of the two games are identical.

Figure 19.2. Game session

Revealing social justice issues

Aware that project development may fail notably because some principles of justice are not considered in the project design, we also proposed an approach to allow the principles of social justice to be revealed upstream of the process of drawing up the community action plans, within the course of three exercises. These exercises were conducted by the researchers with the influential persons. We assumed that the gender issue might be a crucial point in the identification of actions (in the development plan) and should be addressed directly.

In the first exercise, participants were invited to collectively share a situation that they felt was particularly unjust. With the support of a facilitator, a discussion helped to bring out the reasons for the feeling of injustice, which were then reformulated into principles of justice reflecting the group's values (table 19.3). The principles stated here were mainly related to distributive justice, i.e. the final distribution of benefits or losses amongst actors (e.g. unequal access, respect for traditional values) or procedural justice (e.g. discrimination, transparency).

Based on these principles, a second exercise of the JustAGrid type from the CoOPLAGE approach (Ferrand *et al.*, 2017) made the participants aware of the individual variability of the principles of justice and led them to identify the differences between a choice through voting and a choice through consensus.

The third exercise consisted of an anonymous questionnaire to be completed individually. Each participant was asked to indicate a single preference on how funds should be allocated. The result of these exercises highlighted that (i) non-local residents (transhumant or households that moved away) are not considered in the same manner as local resident households, (ii) men and women do not have the same preferences. Women tend to favour a strictly equal allocation or one that favours the most disadvantaged, while men tend to favour a merit-based distribution.

Table 19.3. Principles of justice per sub-group of men and women

Origin of unfair situation for men	Origin of unfair situation for women
Lack of respect: for the public good, for others, for collective rules, for tradition, especially in relation to nature	Lack of respect: for the public good, for others, for collective rules, for tradition, and for commitment
Selfishness and its negative consequences on the group	Selfishness and lack of group spirit and its negative consequences on the community
Discrimination (age, gender)	Lack of love
Unequal access to infrastructure	Lack of transparency in decision-making and corruption
	Unequal situation
	Ignorance

Engaging the non-village stakeholders

The last step of the process was the development of an implementation plan based on the mobilisation of institutionals, NGOs and private actors along with community representatives, in order to obtain a plan negotiated by these different parties. The village authorities were invited to discuss the community action plans, to refine them and to identify legal, financial and organisational constraints and activities that could cause tensions. The aim was also to discuss possible ways of resolving or preventing these tensions, thereby also minimising the risk that the implementation of an action be solely determined by the actor funding it. Indeed, a unilateral approach limits the range of institutional bricolage that is needed at the local level to implement organisational and institutional mechanisms that allow the sustainability of the action.

▶▶ From protocol to implementation: lessons learned

An engaging capacity building process

The capacity process has been developed based on learning-by-doing principles (Kolb *et al.*, 2014). The alternance of collective capacity building sessions and individual on-site implementation was initially conceived as a way to keep with the project planning schedule. It proved to be particularly interesting to support not only the appropriation of the technical aspects of the tools but also to tackle more qualitative posture such as the facilitation posture, strengthen group cohesion as well as build EM confidence in implementation. Thus, this type of training that mixes formal knowledge transfer, mock implementation (in the EM group), immediate implementation and collective exchange of experience facilitates the development of technical skills and functional capacities which help the innovation process to develop (Thoillier *et al.*, 2020).

But the implementation schedule (around three to four weeks to undertake the consultations at village level) was intense and very demanding for the EM who had other responsibilities in the project. Besides, EM had different backgrounds and some of them struggled more than others to master the tools and posture: they would have benefited from a closer on-site coaching. An organisation permitting to have two EM by village during the pilot phase and less time pressure would have increased the benefit of the capacity building. A final evaluation workshop of the complete sequence was also missing.

It is likely that village representatives had also built capacity along the planning process, which could be as important as an identification of priority actions. But this objective had not been sufficiently formalised to really assess how participants benefited from the intensive interactions.

A sectorial and economic view of IWRM

As mentioned, the program emphasised economic valorisation of water resources but excluded "food security" activities from funding. Yet not only flood recession horticulture and fish farming activities are one of the most important livelihood activities locally but they are part of very dynamic value chains as observed in the villages. Moreover, these activities are directly connected to flood plains functioning: in such an environment, IWRM goes beyond issues of multi-use water allocation, water management through supply and demand, or upstream/downstream relationships. Local communities who live in villages surrounded by water six months a year also considered village hygiene and sanitation as key priorities. Yet related activities were funded by another part of the program through projects that intervened in other areas. Thus many priority activities identified during the participatory process could not be financed by the project due to the constraints of the project funder. This situation created frustration for all the actors, facilitators, researchers and representatives of the NGO.

In order to avoid such tensions during the second phase, the NGO asked its local facilitators to guide the discussion around a given set of actions that could be funded: mainly reforestation, riverbank protection, canal cleaning, water hyacinth collection and transformation, and small business. In the pilot villages of the first phase, villagers were encouraged to look for alternative funding opportunities. This adaptive strategy was considered by the program as one of the "good practice" to capitalise on (SNV, 2022). What is at stake here is who should fund and bear the effort to remedy the local impact of water issues, some of them created upstream (nitrogen and phosphorus levels enhancing hyacinth growth) or downstream (urbanisation process driving sand mining for example). But overall, it reveals that the sectorial organisation and utilitarian perspective of the program do not match the way how villagers interact with their environment. A variety of "interests" attach them to their (water) environment, including economy and others. Fostering effective mobilisation to address water issues supposes to consider this diversity of attachment links.

A limited monitoring and evaluation focus that makes the process' assessment difficult

Although we were aware of the importance of Monitoring and Evaluation (M&E), this aspect was clearly not given enough importance during the development of the approach and training of the EM. During the first phase, EM were asked to provide reports at each step of the process following a report template, but they have not received clear recommendations on how to complete this report. Consequently, reporting quality is inconsistent between EM and makes it difficult to really assess the process.

Five out of the 13 M&E reports produced by EM during the first phase permitted a quantitative monitoring of participation. The case reports show that the methodology mobilised between six and 60% of village households. The 'influential' women remained strongly mobilised throughout the process, showing their interest in an approach that allowed them to express themselves independently.

The NGO for itself organised the M&E to monitor activities progress and to comply with the quantitative requirement of the program (surface reclaimed by action, number of women engaged to new value chain activity, number of proto water committee created, number of technical training). The OMIDelta program nonetheless organised sessions of exchange of experiences between projects where more qualitative issues were discussed, such as the difficulty to engage with non-village actors and to institutionalise activities, the tensions linked to land tenure issues, etc.

It is in practice complicated to even know what activity of the process was effectively carried out during the second stage and how they were conducted. An internship work underlined however that (1) implementation plans were not carried out in any of the villages, (2) the process was simplified in the second phase, no workshops dedicated to detailed characterisation of actions were undertaken, and the general discussion of each plan with community members was not conducted either. As a result, the obtained plans are not as substantial as those from the first phase. The simplification was linked to a choice to focus on quick and targeted consultations which better suited the project's agenda and limited travel possibilities due to seasonal constraints. Yet the participatory elaboration of local water management plans is viewed as a key contribution of the project at program level (SNV, 2022).

In the end, a greater number of issues were mentioned in the villages of the second group upon the implementation of actions. Although the level of mobilisation was locally high in the second group, the mobilisation difficulties are more noticeable in the latter than in the first (Yabi, 2021).

Games and ES framework

The game was designed as a tool to introduce ES notions but does not fully engage in ES conceptual framework and its challenges such as how to address trade-offs and synergies of ES and scale emergence issues. Our ambition with the game was not only to support the choice of actions valuing some ES services, but more importantly to engage villagers in discussing the role of the socio-political institutions needed to mobilise these services in practices (Maris, 2014) and thus to discuss the constraints of action implementation. Ultimately, the ES framework was not mobilised in the planning process.

For the NGO, the main interests of the process were the strong feeling of trust between the project and the communities, which was built on the experience of the game and its participatory and mobilising scope for identifying action. Past experience shows that the game does not have a reflective function in essence. Indeed, it is only a tool at the service of a participatory process, the purpose of which may be to encourage discussion, lead to consultation, emancipation or, on the contrary, manipulation. It is therefore extremely important to be aware of this and to underline the importance of training game facilitators on the facilitation posture itself.

A too limited engagement into socio-political issues

Socio-political issues were integrated in the process in an explicit manner in two steps of the process: in the implementation plan and the identification of equity perception. Although the NGO wanted to involve existing institutions in the process, they initially had no clear view of their role in the village activities. We intended the implementation plan to be multi-institutional so as to take into account possible obstacles (land, institutional or legal issues), deal with them and facilitate the institutionalisation process. Yet, the multi-institutional factor and the way technical expertise could intervene was not sufficiently clarified during the training phase. In practice, technical expertise was not really mobilised during the experimental implementation planning process and the NGO assessed the outcomes as unrealistic and technically unsound. The NGO therefore chose to propose their own implementation protocols rather than co-construct them in a multi-institutional approach. On the other hand, the NGO was not experienced enough to anticipate socio- political issues notably those associated with resource tenure that would likely have emerged in a multi-institutional process. Indeed, land conflicts and collaboration of local authorities not surprisingly emerged as some of the key limiting factors to the program (SNV, 2021).

Equity perception exercise was not fully integrated in the participatory process. This was all the more an issue that the choice to rely mainly on influential people is likely to have biased the information gathered even if we sought to have a representation of the different groups. For example, sand mining was discarded by participants because it was considered as a too demanding and dangerous activity but a better representation of less favoured households with more limited livelihood options might have not led to this outcome. The co-option process may also have generated frustrations that the young EM didn't have the time to take into account, or didn't dare to bring to the attention of the NGO leadership.

▸▸ Conclusion

The experience highlights the difficulty of initiating innovative processes in face of the way development programmes work and the expectations of communities with regard to external interventions. The CoOPLAGE "ideal" methodology (if it has ever existed) was implemented under constraints that were imposed on us as experts, but also on the NGO as the operator of a development programme designed by the donor. A number of limitations have been mentioned in the preceding sections. Our aim here was to take stock of them, to reveal the biases that were introduced, and the adaptations that we tried to make to preserve the principles of our approach while trying to respond to the constraints of the NGO. It is a balancing act that we have undertaken.

From the point of view of this book and its didactic ambition, our aim is not to hide them so that, faced with such a situation, other consultants in charge of setting up an IWRM project can anticipate some of these difficulties, and so that donors can propose funding frameworks that are more consistent with the well-known difficulties of setting up IWRM.

▸▸ References

Barreteau O., Antona M., D'Aquino P., Aubert S., Boissau S., Bousquet F., *et al.*, 2003. Our companion modelling approach. *Journal of Artificial Societies and Social Simulation*, 6 (2):1, n.p. http://jasss.soc.surrey.ac.uk/6/2/1.html

Batchelor S., McKemey K., Scott N., 2000. *Exit strategies for resettlement of drought prone populations. Project technical report*, In, 123. London: ODI.

Booth D., 2012. *Development as a collective action problem*, Africa power and politics programme.

Cleaver F., 2012. *Development through bricolage: rethinking institutions for natural resource management*, Routledge.

Ferrand N., Abrami G., Hassenforder E., Noury B., Ducrot R., Farolfi S., *et al.*, 2017. *Coupling for Coping, CoOPLAaGE: an integrative strategy and toolbox fostering multi-level hydrosocial adaptation.* European Union.

Kolb D.A., Boyatzis R.E., Mainemelis C., 2014. Experiential learning theory: Previous research and new directions. *In: Perspectives on thinking, learning, and cognitive styles.* Routledge, p. 227-247.

Maris V. 2014. Nature à vendre : Les limites des services écosystémiques. *Nature à vendre*, 1-96.

Merrey D.J., Cook S., 2012. Fostering institutional creativity at multiple levels: Towards facilitated institutional bricolage. *Water Alternatives*, 5: 1-19.

Millennium Ecosystem Assessment. 2005. *Ecosystems and Human Well-being: Synthesis.* Washington, DC.

Toillier A., Guillonnet R., Bucciarelli M., Hawkins R., 2020. *Developing capacities for agricultural innovation systems: lessons from implementing a common framework in eight countries.* Rome, FAO and Paris, Agrinatura. https://doi.org/10.4060/cb1251en.

SNV, 2021. *Capitalisation de la durabilité : expérience de mise en œuvre de 11 projets AEPHA et GIRE.* OmiDelta, Fond Acteurs Non étatiques, capitalisation des experiences de mise en oeuvre de la durabilité. Rapport, 65 p + Annexes. https://a.storyblok.com/f/191310/c648a7a5fc/2021omideltaane_gire_capitalisationdesexp-c3-a9riences2020-2021_snv_vf.pdf

SNV, 2022. *Six études de cas tirées de la mise en oeuvre des 13 projets financés par le Fonds Acteurs Non Etatiques du Programme OmiDelta.* Rapport. Etude de cas– success stories. 79p. https://a.storyblok.com/f/191310/0b340f21f0/2021omideltaane_successstories_6-c3-a9tudesdecas_snv_vf.pdf

Yabi A., 2021. *Suivi évaluation de l'impact de la démarche participative de valorisation des services écosystémiques liés à la ressource en eau dans la BMVO.* Msc AgroCampus Ouest, 67 p + annexes.

Conclusion

Emeline Hassenforder and Nils Ferrand

This book highlights a multiplicity of pathways, i.e. ways of thinking and implementing participatory approaches in a perspective of sustainability of socio-ecosystems. The book focuses on three main elements that constitute the core of the CoOPLAGE pathways: (self-)engineering and evaluation of participatory processes, modelling and simulation, and planning. These pathways are ongoing: there are as many instances of each tool presented in this book as there are territories where the tool has been implemented. Our research is made and shaped by the socio-ecological system on and for which it works. It is truly stakeholder driven.

Other tools have also been developed before the publication of this book but could not be included.

JustAGrid (2004) is a simple transferable method for eliciting, discussing and choosing principles for shared resources' allocation. It is based on the assumptions that:
— transformative processes in socio-ecological systems redistribute resources' access and use among stakeholders,
— conditions and procedures for this redistribution define social justice in policy design and implementation and,
— stakeholders who can't elicit, deliberate or choose these are at least exposed to unfair treatment (in democratic terms).

Hence a tool for participants to first decide individually of resources' allocation, followed by an aggregation and a collective discussion toward a common proposal of joint principles for shared resources' allocation.

SMAG (2016) for Self-Modelling for Assessing Governance, is a collaborative tool aimed at letting some selected stakeholders remapping the past evolution of governance, through its impact on space, and the most important decisions taken, driven by estimated causes, with induced consequences. The role of local stakeholders is elicited. Conclusions build on this past analysis to propose adaptation of the governance patterns.

Training and MOOC: the CoOPLAGE tools are regularly the subject of initial and professional training. More than 3,500 persons, students and professionals, have been trained (2023). A comprehensive MOOC (eight modules, ~50h training) is available since 2019 that accompanies students to collectively develop their own "CoOPLAGE pathways".

CoOPLANET network: many CoOPLAGE contributors and users are part of a network called "CoOPLANET" intended to share experiences among its members, and validate their expertise. It was launched at the COP22 in Marrakech (2016).

Additionally, a number of new tools have been developed during and since the writing of this book, all aimed at guiding stakeholder participation in the decision-making process toward socio-ecological sustainability (figure 19.1). They open new and alternative pathways that we have begun to explore in response to the demands of stakeholders in the field.

The River Observation and Conservation Kit (ROCK) allows citizens to define what they want and need to know about water or land management, why, how they can get this information and from whom. In other words, ROCK allows to frame informational needs and services, and thus contributes to establish (in a participatory way) a participatory observation device. With ROCK, it is not only the researchers, but also the citizens, who define what information to collect, with whom, where and how. ROCK is materialised in a two-sided form that leads participants to question their information needs and how to respond to them (figure 19.1).

ChangeO'Log is a tool aiming at exploring collectively various change or preservation pathways toward socio-ecological sustainability. It combines PrePar, CoOPLAN and ENCORE described in this book. Participants collectively define what they think should be changed, preserved or adapted in the socio-ecological system. They identify the actors who should act to achieve the proposed changes and those who would be impacted. They identify the management, participation and monitoring and evaluation

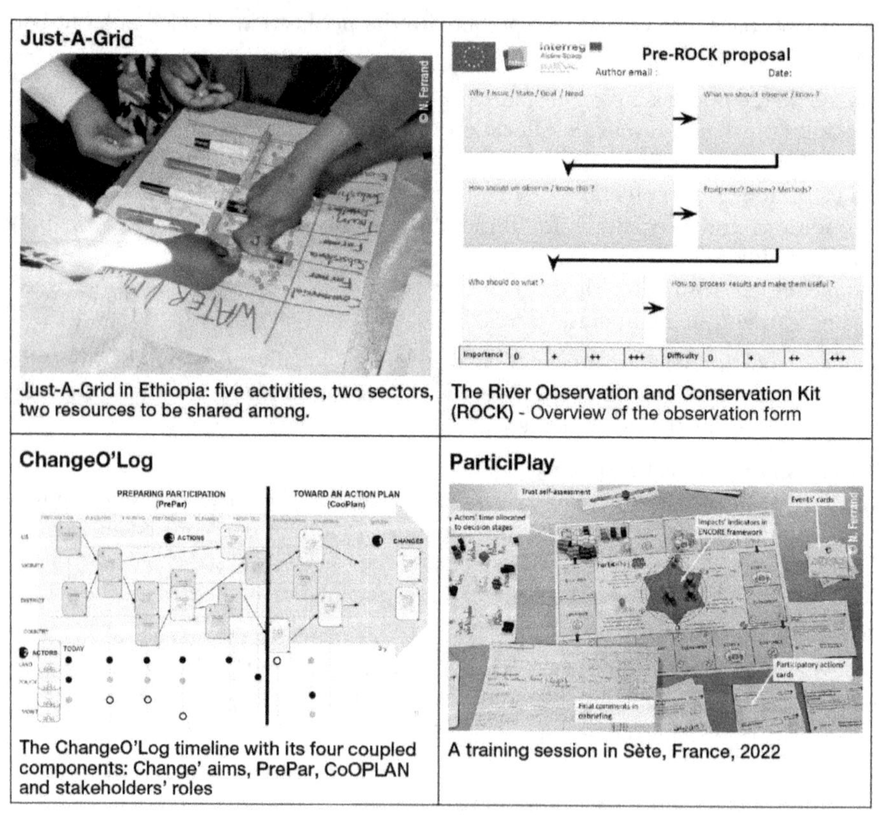

Just-A-Grid in Ethiopia: five activities, two sectors, two resources to be shared among.	The River Observation and Conservation Kit (ROCK) - Overview of the observation form
The ChangeO'Log timeline with its four coupled components: Change' aims, PrePar, CoOPLAN and stakeholders' roles	A training session in Sète, France, 2022

Figure C.1. Four other developments in the CoOPLAGE suite of tools actions to be implemented to achieve the desired changes. And they articulate these elements in a coherent whole discussed with all the actors concerned. ChangeO'Log therefore brings together participation engineering, planning and monitoring-evaluation of the desired changes towards socio-ecological sustainability.

Participlay is a simple and accessible game aiming at initiating and quickly engaging all audiences to the issues and practices of participatory decision-making and open innovation. Participants collectively set up "participation action" cards, face unexpected events, discuss the social impacts of participation, orient themselves among different trajectories and possible approaches, and mobilise collective intelligence and creativity for innovation.

⯈ Pending issues and way forward

Within this community, several other issues have emerged over time, and we couldn't address all in this book. Some appear nowadays to be increasingly urgent to address:

Social extension, massification and adoption for general public policies: most of CoOPLAGE developments and tools have emerged in projects led by researchers in collaboration with public institutions. The social pervasiveness is still limited. It's time to foster more "massive" processes, still in line with the principles, but able to cope with very large groups, become prevalent in day-to-day approaches of public affairs, and even being adopted as a "normal" procedure in public policy making and implementation. Online and mobile apps may help, but citizens' engagement winning strategies are still to build. Improvement of guidelines and communication is urgent.

Reassessing long term impacts: some of these applied projects are 15 years old. When transformative impacts were measured, it was soon after the funded phase. It's time to come back and reassess the long-term impacts, in the various terms of the ENCORE framework. Causal imputability may be a challenge, but the perception of stakeholders and their vision of the remnants, when triangulated, could be useful.

Enriching CoOPLAN with empirical evaluative models: in participatory planning with CoOPLAN, the cross-evaluation of the integrated plan with a qualitative and deliberative appraisal (for resources and impacts) is very arguable and contingent. In collaboration with the French unit INRIA STEEP specialised in Material Flow Analysis models (territorial metabolism – Courtonne *et al.*, 2015), or with other categories of socio-environmental models (e.g. multi-agent) we started recoupling the set of actions with an empirical model to compute directly some indicators of impact, to feed the deliberation.

Extending digital participation engineering with CoOPILOT: the development of CoOPILOT (chapter 8) has been a long process. It transferred the knowledge and tools of the CoOPLAGE community to generate a significant breakthrough in support to participation engineering, by integrating new tools like ePrePar. However, its extension for public policy makers and administrators is still pending. As for the physical CoOPLAGE, it may require several adaptations and specialisation, or marketing efforts.

Migrating to other application domains: Developed with a focal issue on integrated water management, CoOPLAGE has, through field applications, addressed a very large set of other domains (risks, biodiversity, energy, food, health, urban, transitions, SDGs). However, some generalisations and transfers are still to be formalised and achieved, as well as improved scientific collaborations with these domains' specialists.

Dealing with emotions and affects – linking with art: the CoOPLAGe approaches are in line with operational research, hence quite "dry" for accounting for emotions and affects, which we reckon to drive several decisions and behaviors. Although engaging the "real" humans in the process, it doesn't give much space and incentive for expressing, sharing and valuing emotions. Participatory modelling and role-playing games offer some. Justice dialogues also trigger "outing" of emotions. However, many alternative combinations or add-on could be considered, with a special attention to be paid to the use of artistic movements. With various media, involving participants in gestures not directly "useful" or productive, can strengthen commitment and social learning. The role of artists themselves could also be reinforced.

Linking with prospective thinking and anticipation: Given the rapid evolution of socio-ecological systems, anticipation, understood as an effort to "know" the future, in the sense of "thinking the future" and "using the future" (Miller *et al.*, 2018) appears as a natural perspective for coupling with CoOPLAGE tools. The objective is to develop stakeholder capacities to anticipate sudden shifts and shocks and to make decisions accounting for uncertainty and unpredictability (Rutting *et al.*, 2022). Several such experiments have already been implemented, such as the modelling of past, current and future groundwater governance in Tunisia, using visioning and the Futures Triangle.

Restructuring the role of scientific and technical expertise: in the CoOPLAGE tools, scientific and technical experts of the application domains play a limited role. For participatory modelling and planning, their intervention could be normalised when reassessing actions' needs and impacts, as well as when analysing the global action plan as a whole. However, there are methodological challenges when confronting expert protocols and models with the simplified but pragmatic and open formalisms of the CoOPLAGE tools.

▸▸ Calling for future collaboration

CoOPLAGE is almost entirely open source, under Creative Commons by-nc-sa license, with a specific use agreement including a protocol for sharing the monitoring and evaluation results.

In such context, we encourage all researchers, practitioners or other stakeholders to reuse the tools, share the results, adapt the methodology and contribute to the community. The core CoOPLAGE group is very tiny, but willing to help and establish common new projects for exploring new issues, domains, tools or challenges. Let's share the best of participation engineering to support the urgent transitions our contemporary world require!

▸▸ References

Courtonne J.Y., Alapetite J., Longaretti P.Y., *et al.*, 2015. Downscaling material flow analysis: The case of the cereal supply chain in France, *Ecological Economics*, 118, 67–80.

Miller R., Poli R., Rossel P., 2018. The discipline of anticipation: Foundations for futures literacy. *In:* Miller R. (ed), *Transforming the future, Anticipation in the 21st Century*, Routledge, 51-65.

Rutting L., Vervoort J., Mees H., Driessen P., 2022. Strengthening foresight for governance of social-ecological systems: An interdisciplinary perspective. *Futures, 141*(102988). https://doi.org/10.1016/j.futures.2022.102988

List of abbreviations

ACTED: A French international NGO acting in development and crisis management

AERMC: French Water Agency from Rhône-Mediterrannée-Corse Catchment

AFD: Agence Française de Développement – French Development Agency

Capp'WAG: a game apparatus devoted to the assessment of collective capabilities through compared game sessions

CBO: Community-Based Organisation

Cerema: Centre d'études et d'expertise sur les risques, l'environnement, la mobilité et l'aménagement – Centre for Studies and Expertise on Risks, the Environment, Mobility and Urban Planning

ChangeO'Log: a participatory apparatus to design and support change processes

Cirad: Centre de coopération internationale en recherche agronomique pour le développement – French Agricultural Research Centre for International Development

CNDP: French National Commission for Public Debate

ComMod: Companion Modelling

CoOPILOT: name of the digital platform containing all the CoOPLAGE tools

CoOPLAGE: Coupler des outils ouverts et participatifs pour laisser les acteurs s'adapter pour la gestion de l'environnement – Coupling Open and Participatory Tools to Let Actors Adapt for Environmental Management

CoOPLAN: name of the CoOPLAGE tool aimed at participatory planning

CoOPLANET: an international network of practitioners of CoOPLAGE

Cormas: COmmon pool Ressources and Multi-Agent Simulations platform

COSTEA: Comité Scientifique et technique de l'Eau Agricole – Scientific and Technical Committee for Agricultural Water

CreaWAG: a method for getting stakeholders to create their own Wat-A-Game model

CSO: Civil Society Organisation

DCCEEW: Department of Climate Change, Energy, the Environment and Water (Australian Government)

DoUbT: Deltas' Dealings with Uncertainty project

ENCORE: External, Normative, Cognitive, Operational, Relational, Equity (the different types of impacts that can be evaluated)

FFEM: Fonds français pour l'environnement mondial – French Facility for Global Environment

G-EAU: Joint research unit "Water Matters" in Montpellier

GIS: Geographical information system

ICT: Information and communication technologies

IEA: Institution of Engineers Australia

Igref: Ingénieurs du génie rural, des eaux et des forêts – Rural, water and forestry engineers –

INI-WAG: Wat-A-Game basic kit to understand the principles of an integrated water management role-play

INRAE: Institut national de recherche pour l'agriculture, l'alimentation et l'environnement – French National Research Institute for Agriculture, Food and Environment

Ipef: Ingénieurs des ponts, des eaux et des forêts – Bridges, water and forestry engineers

IRD: Institut de recherche pour le développement – French Research Institute for Development

ISAGA: International Simulation and Gaming Conference

ISC: Irrigation Service Center (Non-governmental organisation, Cambodia)

IWRM: Integrated Water Resource Management – GIRE in French for Gestion Intégrée des Ressources en Eau

JustAGrid: name of the CoOPLAGE tool aimed at supporting dialogue on distributive justice principles

LittoSIM: a digital apparatus dealing with coastal area management

LittoWAG: name of the CoOPLAGE tool on coastal area management and seaside adaptation (based on Wat-A-Game)

M&E: Monitoring and Evaluation

MISE: Inter-Service Mission on Water

MOOC: Massive Online Open Course

MoWRAM: Ministry of Water Resources and Meteorology, Cambodia

Mpan'Game: name of the CoOPLAGE tool dealing with management of the Mpanga river in western Uganda (based on Wat-A-Game)

MUSE KIM WATERS: Key Initiative on Water from Montpellier University

MyRiverKit: name of the CoOPLAGE tool dealing with on river ecosystem services and management (based on Wat-A-Game)

NGO: Non-governmental organisation

OECD: Organisation for Economic Cooperation and Development

PACTE: Climate Change Adaptation Programme for Vulnerable Rural Territories

ParticiPlay: name of the CoOPLAGE tool aimed at initiating and quickly engaging all audiences to the issues and practices of participatory decision-making and open innovation

PEP: Politique de l'Eau Partagée – Shared Water Policy from New-Caledonia government

PLANISSIM: a European project (Europ-Aid) dealing with wastewater and sanitation management in Senegalese communities

PrePar: name of the CoOPLAGE tool aimed at preparing/engineering a participatory process

RCAD: Tunisian Regional Commission for Agricultural Development

RDO: Rural Development Officer

ROCK: River Observation and Conservation Kit, name of the CoOPLAGE tool aimed at participatory assessment of observation and knowledge needs

RUA: Royal University of Agriculture, Cambodia

SAGE: Schéma d'aménagement et de gestion des eaux – Water Management Scheme

Sertoes: Sustainability and Water Resilience of Territories in the Northeast Brazil (project)

SMAG: Self-Modelling for Assessing Governance

SMBVT: Syndicat Mixte du Bassin versant de la Têt

SPARE: Strategic Planning for Alpine River Ecosystems (European project)

Terr'Eau & co: a MOOC on the CoOPLAGE tools

WAG: Wat-A-Game

WasteWAG: wastewater game (based on Wat-A-Game)

WAT4CAM: Water Resources Management & Agricultural Transition for Cambodia Project

Acknowledgments

Chapter 1. The author acknowledges the following current funding sources which have supported the research and perspectives underlying this chapter: ANU Futures grant, a Jean Monnet EC Project grant for the "Algorithmic Futures Policy Lab" and a DSTG Philosophy of Operations Research grant. Initial work with the developers of CoOPLAGE was supported by a John Monash Scholarship and a range of other institutions, including the French Ministry of Foreign Affairs through a "Fonds Pacifique" grant. Thank you to all my colleagues who have been a part of building and reflecting on these participatory methods in many places and projects. A particular acknowledgement to Flynn Shaw, ANU School of Cybernetics and Fenner School of Environment and Society, for his research assistance on this chapter.

Chapter 3. This work was carried out by the French National Institute for Research in Science and Technology for the Environment and Agriculture (Irstea) for the Rhône Méditerranée Corse Water Agency. It is part of a framework agreement between the two parties for a three-and-a-half year project (mid 2016 to end 2019) entitled "Which participatory strategy for local water management with citizens?"

Chapter 5. This work was supported by the "Climate Change Adaptation Program for Vulnerable Rural Territories of Tunisia" (PACTE, 2018-2024), funded by the the Agence Française de Développement (AFD), the Fonds Français pour l'Environnement Mondial (FFEM) and the Republic of Tunisia.

Chapter 6. Work on Case no.1 (Water reuse) was supported by the Read'Apt project (2016-2019) "Réutilisation des eaux usées en agriculture dans une approche de projet de territoire" funded by the French Water Agency Rhône-Méditerranée-Corse (Action program 2013-2018). Work on Case no.2 (Artificial wetland buffer zones) was supported by the Brie'eau project (2016-2020) "Vers une nouvelle construction de paysage agricole et écologique sur le territoire de la Brie : associer qualité de l'eau et biodiversité" funded by the Partnership research program "Pour et Sur le Développement Régional 4" (PSDR 4) Ile-de-France, and the Interdisciplinary Research Program on Water and the Environment in the Seine Basin (Piren-Seine).

Chapter 7. The work in Tunisia was supported by the "Climate Change Adaptation Program for Vulnerable Rural Territories of Tunisia" (PACTE, 2018-2024), funded by the the Agence Française de Développement (AFD), the Fonds Français pour l'Environnement Mondial (FFEM) and the Republic of Tunisia. The work in Brazil was supported by the Sertoes project (Sustainability and Water Resilience of Territories in the Northeast Brazil, 2020-2023), funded by the the Agence Française de Développement (AFD) through the Facilité 2050 (grant agreement number PAR-MOA-001).

Chapter 8. The design and development of the e-CoOPILOT platform was shared between Resurgences R&D, now Ananke, a French R&D company from the social economy sector, and IRSTEA, now INRAE, with the additional support of the

INTERREG Alpine Space SPARE project (2014-2020) and the Programme Investisse-ment d'Avenir – Territoire Innovant de Grande Ambition "Littoral+", with Région Occitanie and Banque des Territoires (2020-2030). The "Terr'Eau & co" MOOC devel-opment was funded under the project AgreenCamp of the program IDEFI-N of the French National Research Agency (2016-2020).

Chapter 9. The PrePar methodology has been evolved through various European projects: HarmoniCOP (2002-2005 - EU FP5), Aquastress (2005-2009 – EU FP6) and the INTERREG Alpine Space SPARE project (2014-2020). Some developments were based on a consultancy for the UN General Secretary – DESA-DSD for the UN Office of Sustainable Development (Atkison, Ferrand, 2011).

Chapter 10. The ENCORE evaluation framework was initiated in a research action funded under the "Concertation Décision en Environnement" French program (2004), followed by developments in the HarmoniCA concerted action (EU FP6 EVK1-2001-00192), and further extensions in the Aquastress (2005-2009 – EU FP6) and the Afromaison (2011-2014 – EUFP7) projects.

Chapter 11. Work on example 1 was carried out during a PhD funded by the French Ministry of Education and Research. Work on example 2 was supported by the Brie'eau project (2016-2020), "Vers une nouvelle construction de paysage agricole et écologique sur le territoire de la Brie : associer qualité de l'eau et biodiversité" funded by the Partnership research program "Pour et Sur le Développement Régional 4" (PSDR 4) Ile-de-France, and the Interdisciplinary Research Program on Water and the Environment in the Seine Basin (Piren-Seine). Work on example 3 in Drôme was supported by Project "SPARE - Strategic Planning for Alpine River Ecosystems" (2014-2020) co-financed by the European Regional Development Fund through the Alpine Space Programme (2014 – 2020 INTERREG VB ALPINE SPACE). Work on example 4 in Drôme and Cèze was funded by the projet "Apprendre la rareté de l'eau sur un territoire, pour aujourd'hui et demain" ("Learning about water scarcity in a territory, for today and tomorrow") which is part of the 2016 framework agreement between the Rhône-Méditerranée-Corse Water Agency and the Zone atelier du Bassin du Rhône (ZABR, Rhône Basin Workshop Zone). Work on example 5 was carried out during a PhD funded by the French Doctoral School on Agriculture Food Biology Environment Health (ABIES) and by the former French National research institute of science and technology for environment and agriculture (ex-Irstea, now INRAE).

Chapter 13. This work was supported by the Water Agency of Rhône-Méditer-ranée-Corse ("AAP lycées agricoles" 2014-2017, N°513.2014-090-0SA and 2017-2020, N°513.2017.142-OSK), the Regional Council Languedoc-Roussillon (2014-2017, grant agreement N° 2014 004953), the Regional Council Occitanie Pyrénées-Méditerranée (2018-2020, grant agreement N°2018 002185) and the French Ministry of Agriculture, the CGED (General Commissariat for the Equality of Territories) and the European Fund for Agriculture and Rural Development (specific programm of "Réseau Rural National 2014-2020 : Valorisation and sharing of projects concerning water resource protection in french agricultural learning system", N°RRRN200219NA1000006).

Chapter 14. This work was supported by the PLANISSIM project (2017-2018) "Sani-tation planing through participatory modelling and simulation" funded through the 10th PAISC (civil society support program) of the Senegales Government by the European Developement Fund "FED/2017" grant agreement number 385-740.

Chapter 15. This research has been conducted in the framework of the Deltas' Dealings with Uncertainty (DoUbT) and COSTEA projects. Financial support from the French National Agency for Research (ANR) and the Japan Society for the Promotion of Science (JPSP) in the context of the Open Research Area (ORA) for the Social Sciences program, and from the French Agency of Development (AFD) through the Association Française de l'Eau, l'Irrigation et le Drainage (AFEID) and the COSTEA programme is duly acknowledged.

Chapter 16. This work was supported by the Key Initiative MUSE: Montpellier University project (2018-2022), funded by the isite programme of the French National Research Agency, grant agreement number ANR-16-IDEX-0006. Following the project presented in the chapter, so-ii was officially recognised and launched in December 2019 when it was labeled as an observation system of OSU-OREME (Mediterranean Environmental Research Observatory).

Chapter 17. The work in Uganda was supported by the AfroMaison project (2011-2014) "Africa at a meso-scale: Adaptive and integrated tools and strategies for natural resource management" funded by the 7th Framework Program of the European Union, theme "ENV.2010.2.1.1-1" [Integrated management of water and other natural resources in Africa], grant agreement number 266379. The work in France was supported by the Project "SPARE - Strategic Planning for Alpine River Ecosystems" (2014-2020) co-financed by the European Regional Development Fund through the Alpine Space Programme (2014 – 2020 INTERREG VB ALPINE SPACE). The work in Tunisia was supported by the "Climate Change Adaptation Program for Vulnerable Rural Territories of Tunisia" (PACTE, 2018-2024), funded by the the Agence Française de Développement (AFD), the Fonds Français pour l'Environnement Mondial (FFEM) and the Republic of Tunisia.

Chapter 19. This work was developed by Cirad researchers under the responsibility of ANTEA as part of the NVW-GIRE (Nouvelle Vallée de l'Ouémé GIRE) project coordinated by Join For Water. The project was funded by the Non-State Actors Fund of the OmiDelta program of the Dutch Embassy in Benin.

List of authors

Géraldine ABRAMI: G-EAU, Univ Montpellier, AgroParisTech, BRGM, CIRAD, INRAE, Institut Agro, IRD, Montpellier, France.

Alexandre ALIX: French Water Partnership, France.

Mélaine AUCANTE: Loire-Bretagne Water Agency, France.

Alpha BA: University of Thiès, National School of Agriculture (ENSA), Department of Economics and Rural Sociology, Senegal.

David BADOGA: Ngaoundéré University, Faculty of Sciences, Department of Earth Sciences, Cameroon.

Pierre BALZERGUE: G-EAU, Univ Montpellier, AgroParisTech, BRGM, CIRAD, INRAE, Institut Agro, IRD, Montpellier, France.

Audrey BARBE: Lisode, France.

Nicolas BECU: French National Centre for Scientific Research (CNRS), joint research unit on LIttoral, ENvironment and Societies (UMR LIENSs), France.

Anissa BEN HASSINE: Bizerte Regional Agricultural Development Commission, Tunisia.

Valérie BORRELL ESTUPINA: University of Montpellier, joint research unit "Water Matters" (UMR G-EAU), France.

Séverine BOUARD: New Caledonian Institute of Agronomy (IAC), Research Center North Thierry Mennesson, New Caledonia.

Houssem BRAÏKI: Independent consultant.

Pauline BRÉMOND: G-EAU, Univ Montpellier, AgroParisTech, BRGM, CIRAD, INRAE, Institut Agro, IRD, Montpellier, France.

Julien BURTE: CIRAD, UMR G-EAU, 6202 Rabat, Morocco. G-EAU, Univ Montpellier, AgroParisTech, CIRAD, INRAE, Institut Agro, IRD, Montpellier, France.

Fabrice CAROL: Mixed syndicate for the Têt river basin, France.

Camille CHEVAL: Expertise France, France.

Dominique DALBIN: Réso'them, Ministry of Agriculture and Food Sovereignty, France.

Katherine Anne DANIELL: The Australian National University (ANU), School of Cybernetics, Fenner School of Environment and Society, Institute for Water Futures, Australia.

William's DARÉ: CIRAD, UMR SENS, F-34398 Montpellier, France. SENS, CIRAD, IRD, Université de Paul Valéry Montpellier 3, Montpellier, France.

Etienne DELAY: CIRAD, UMR SENS, Dakar, Sénégal. SENS, CIRAD, IRD, Université de Paul Valéry Montpellier 3, Montpellier, France.

Moustapha DJANGUE: Ngaoundéré University, Faculty of Sciences, Department of Earth Sciences, Cameroon.

Raphaëlle DUCROT: CIRAD, UMR G-EAU, 2696 Phnom Penh, Cambodia. G-EAU, Univ Montpellier, AgroParisTech, CIRAD, INRAE, Institut Agro, IRD, Montpellier, France.

Chrystel FERMOND: Mixed syndicate for the Drôme river and its tributaries (SMRD), France.

Nils FERRAND: G-EAU, Univ Montpellier, AgroParisTech, BRGM, CIRAD, INRAE, Institut Agro, IRD, Montpellier, France. French National Research Institute for Digital Science and Technology (INRIA).

Alain FEZEU: French National Research Institute on Development (IRD), Cameroon.

Patrice GARIN: G-EAU, Univ Montpellier, AgroParisTech, BRGM, CIRAD, INRAE, Institut Agro, IRD, Montpellier, France.

Sabine GIRARD: French National Research Institute on Agriculture, Food and Environment (INRAE), Mountain Ecosystems and Societies Laboratory (LESSEM), France.

Frédéric GRELOT: G-EAU, Univ Montpellier, AgroParisTech, BRGM, CIRAD, INRAE, Institut Agro, IRD, Montpellier, France.

Fethi HADAJI: Kairouan Regional Agricultural Development Commission, Tunisia.

Mohamed Chamseddine HARRABI: Ministry of Agriculture, Hydraulic Resources and Fisheries (MARHP), General Direction for the Development and Conservation of Agricultural Land (DG-ACTA), Tunisia.

Emeline HASSENFORDER: CIRAD, UMR G-EAU, 01800 Tunis, Tunisia. G-EAU, Univ Montpellier, AgroParisTech, CIRAD, INRAE, Institut Agro, IRD, Montpellier, France.

Sreytouch HEOURN: Royal University of Agriculture, Cambodia.

Jofre HERRERO FERRAN: Water, Air and Soil Unit, Eurecat—Technological Centre of Catalonia, Spain, and University of Barcelona, Spain.

Anne HILLERET: French Centre for Studies and Expertise on Risks, the Environment, Mobility and Urban Planning (Cerema), France.

Amar IMACHE: Lisode, France.

Joana JANIW: General Commission for Sustainable Development, Ministry of Ecological Transition and Territorial Cohesion, Ministry of Energy Transition, France.

Camille JOURDAN: BRL Ingénierie (BRLi), France.

Clovis KABASEKE: Mountains of the Moon University, Uganda.

Karine LANCEMENT: French Centre for Studies and Expertise on Risks, the Environment, Mobility and Urban Planning (Cerema), France.

Julie LATUNE: G-EAU, Univ Montpellier, AgroParisTech, BRGM, CIRAD, INRAE, Institut Agro, IRD, Montpellier, France.

Floriane LE MOING: Mixed syndicate for the Têt river basin, France.

Caroline LEJARS: CIRAD, UMR G-EAU, F-34398 Montpellier, France. G-EAU, Univ Montpellier, AgroParisTech, CIRAD, INRAE, Institut Agro, IRD, Montpellier, France.

Guillaume LESTRELIN: CIRAD, UMR TETIS, 1082 Tunis, Tunisie. TETIS, Univ Montpellier, AgroParisTech, CIRAD, CNRS, INRAE, Montpellier, France.

Cécile LLOVEL: Philia Ingénierie, France.

Rémi LOMBARD-LATUNE: French National Research Institute on Agriculture, Food and Environment (INRAE), Research unit "Reduce, Reuse, and Recover Wastewater Resources" (UR REVERSAAL), France.

Lamphin LOR: Irrigation Service Center, Cambodia.

Sarah LOUDIN: ACTeon, France.

Linda LUQUOT: Géosciences Montpellier, French National Centre for Scientific Research (CNRS), University of Montpellier, France.

Valérianne MARRY: G-EAU, Univ Montpellier, AgroParisTech, BRGM, CIRAD, INRAE, Institut Agro, IRD, Montpellier, France.

Audrey MASSOT: Ministry of Ecological Transition and Territorial Cohesion, Water and Biodiversity Directorate, France.

Veronica MITROI: CIRAD, UMR G-EAU, 60115-221 Fortaleza CE, Brasil. G-EAU, Univ Montpellier, AgroParisTech, CIRAD, INRAE, Institut Agro, IRD, Montpellier, France.

Sylvie MORARDET: G-EAU, Univ Montpellier, AgroParisTech, BRGM, CIRAD, INRAE, Institut Agro, IRD, Montpellier, France.

Paul MORETTI: Cabinet Merlin, France.

Roger MOUSSA: French National Research Institute on Agriculture, Food and Environment (INRAE), Laboratory for the study of Soil-Agrosystem-Hydrosystem interactions (LISAH).

Moses MUHUMUZA: Mountains of the Moon University, Uganda.

Malyne NEANG: Royal University of Agriculture, Cambodia.

Benjamin NOURY: Independent consultant.

Somali OUM: Independent researcher.

Eva PERRIER: G-EAU, Univ Montpellier, AgroParisTech, BRGM, CIRAD, INRAE, Institut Agro, IRD, Montpellier, France.

Claire PETITJEAN: Mixed syndicate for the Drôme river and its tributaries (SMRD), France.

Raksmey PHOEURK: Royal University of Agriculture, Cambodia.

Anne PRESSUROT: Rhone-Mediterranean and Corsica Water Agency, France. Anne.

Diana PUIGSERVER CUERDA: Faculty of Earth Sciences, University of Barcelona, Spain.

Sophie RICHARD: AgroParisTech, joint research unit "Water Matters" (UMR G-EAU), France.

Mariana RIOS: G-EAU, Univ Montpellier, AgroParisTech, BRGM, CIRAD, INRAE, Institut Agro, IRD, Montpellier, France.

Patrice ROBIN: Local public establishment for agricultural education and vocational training (EPLEFPA) Perpignan Roussillon, France.

Marine ROUSSEAU: University of Montpellier, France.

Sopheaktra SAY: Independent consultant.

David SEBAG: IFP Energies nouvelles (IFPEN), Géosciences Direction, France.

Laura SEGUIN: French geological survey (BRGM), joint research unit "Water Matters" (UMR G-EAU), France.

Sophak SENG: Irrigation Service Center, Cambodia.

Eric SERVAT: International Center for Interdisciplinary Research on Water Systems Dynamics, Observatory of Universe Sciences Montpellier Environmental Research Observatory (OSU OREME), France.

Sreypich SINH: French National Research Institute on Development (IRD).

Thaddeo TIBASIIMA: University of Natural Resources and Life Sciences (BOKU), Department of Sustainable Agricultural Systems, Division of Organic Farming, Austria Mountains of the Moon University (MMU), Uganda.

Samuel TRONÇON: Résurgences R&D & SCIC Ananke, France.

Marie TROUILLET: Permanent Center for Environmental Initiatives (CPIE) Bugey Genevois, France.

Sandra VAN-EXTER: Géosciences Montpellier, French National Centre for Scientific Research (CNRS), University of Montpellier, France.

Jean-Philippe VENOT: French National Research Institute on Development (IRD), joint research unit "Water Matters" (UMR G-EAU), France.

Soumaya YOUNSI: Independent consultant.

List of contributors

Since its beginnings in 2008, CoOPLAGE has been continually enriched by many researchers, practitioners and students. In 2010, a fictional author named "Wanda Aquae-Gaudi" was created to represent this composite and interdisciplinary collective. We would like to acknowledge here all the people who are behind Wanda and thank them again for their valuable contribution over the years: Géraldine Abrami, Mathilde Arène, Mélaine Aucante, Xavier Augusseau, Katleena Awan, Etienne Babin, Kahina Baha, Rémi Barbier, Olivier Barreteau, Pavel Bautista, Anaïs Bazi, Syrine Ben Slimane, Jérome Betrancourt, Mathilde Boissier, Adèle Bommier, Mathilde Bonifazi, Sophie Bonnard, Bruno Bonté, Valérie Borrell, Pieter Bots, Séverine Bouard, Sami Bouarfa, Martin Boulay, Perrine Bouteloup, Houssem Braïki, Aglaé Bullich, Julien Burte, Stéphanie Cabantous, Martin Cavero, Eric Cuenca, Beth Cullen, Katherine Daniell, William's Daré, Arnaud De Chonski, Heleen De Fooij, Sander De Waard, Tom Dhaeyer, Mori Diallo, Mathieu Dionnet, Derick Du Toit, Raphaele Ducrot, Karim Ergaieg, Stefano Farolfi, Jérome Fayard, Nils Ferrand, Victor Fleurant, Morgane Ganault, Patrice Garin, Ines Ghazouani, Sabine Girard, Frédéric Grelot, Florestan Groult, Laetitia Guerin, Ernst Gumpinger, Emeline Hassenforder, Thomas Hertzog, Nicolas Huet, Amar Imache, Jean-Yves Jamin, Manuel Joseph-Monrose, Clovis Kabaseke, Sarra Kekli, Julie Latune, Marjorie Le Bars, Nicolas Le Bon, Caroline Lejars, Zelalem Lema, Guillaume Lestrelin, Elsa Leteurtre, André Levy, Rémi Lombard-Latune, Sébastien Loubier, Sarah Loudin, Christine M'Buru, Manuel Magombei, Julien Malard, Maria Mañez-Costa, Ryan McAllistair, Eduardo Martins, Sylvie Morardet, Paul Moretti, Benjamin Noury, Sian Oosthuizen, Eva Perrier, Corentin Piccoli, Marine Pizette, Sharon Pollard, Carmen Renaudeau, Irina Ribarova, Audrey Richard, Mariana Rios, Patrice Robin, Dominique Rollin, Jean-Emmanuel Rougier, Jean-Baptiste Roux, Alyonah Rydannikh, Ardo Samba, Laura Seguin, Caroline Sourzac-Lami, Peter Sturm, Thierno Thiam, Samuel Tronçon, Albena Varsano, Yorck Von Korf, Jade Wolff, Soumaya Younsi, Aida Zare, and all the students of the CoOPLAGE courses.

Layout: Hélène Bonnet Studio 9

Fichier préparé par Nicolas Perrier, société 4P
Imprimé pour vous par Books on Demand (Allemagne)